Systems Analysis for
Information Retrieval

Institute of Information Scientists
657 *High Road, Tottenham, London N17 8AA*
MONOGRAPH SERIES
Hon. General Editor:
Dr D. J. Campbell

In the same series

D. V. ARNOLD, The Management of the Information Department. 1976, pp. 143.

A. H. HOLLOWAY and ELIZABETH H. RIDLER, Information Work with Unpublished Reports. Part I, Work in Large National Information Centres, with supplementary (North American) material by Domenic A. Fuccillo and a section on 'Mechanization' by Marvin E. Wilson. Combined with:
B. YATES, Information Work with Unpublished Reports. Part II, Work in Company-based Information Units. 1976, pp. 302.

Systems Analysis for Information Retrieval

Helen M. Townley
I. R. Advisory Services

A Grafton Book

André Deutsch
IN ASSOCIATION WITH
THE INSTITUTE OF INFORMATION SCIENTISTS

First published 1978 by
André Deutsch Limited
105 Great Russell Street London WC1

Copyright © 1978 by Helen M. Townley
All rights reserved

Printed in Great Britain by
Ebenezer Baylis and Son Limited
The Trinity Press, Worcester, and London

ISBN 0 233 96920 9

Distributed in the United States and Canada
by Westview Press, 1898 Flatiron Court,
Boulder, Colorado 80301, USA.

Dedicated with great affection
to Mary and John Finney,
who did for me
what no one else could do.

Introduction to the Series

The Monograph Series has been planned and organized by the Institute of Information Scientists, in consultation with the publishers. The aim is to provide a series of texts to fill (so far as may be practicable) the considerable gaps in the monograph literature of this fairly new subject, which have greatly complicated the teaching of it; the monographs should be suitable also for people learning on the job, and for information scientists who want to brush up their knowledge of particular fields.

The importance of clear, readable English has been stressed to all authors and, as far as human frailty (my own as well as others') allows, I have tried to insist on it in editing.

The authors are all well-known experts in their fields. Every monograph has been read and criticized by two referees, one of whom is normally a North American and one a Briton, and changes have been made where necessary to meet their comments. The series is intended for English-reading people interested in the subject all over the world.

D. J. CAMPBELL

Contents

	page
List of Figures	xi
Preface	xiii

1. PRELIMINARY IDEAS 1

    1 – What is a system?; 2 – Components of data bases; 4 – The elements of documentary information storage and retrieval (ISR) systems; 5 – The functions of an ISR system; 8 – Varieties of information system requirements; 8 – The subject matter of the base; 10 – The nature of the response; 12 – The 'life' of the data or information; 13 – Response times; effects on design.

2. THE INITIAL SYSTEM STUDY 15

    15 – System designer and information staff; 16 – Statement of goals; 18 – Statement of requirements; 20 – System study; 20 – Inputs; 23 – Records; 25 – Outputs; 25 – Approaching the system study; 28 – Group discussions; 30 – What the analyst suggests.

3. STORAGE: SYSTEMS, FILES, RECORDS AND FIELDS 31

    31 – Access; 31 – Fields; 33 – Records; 39 – Files; 40 – Accessing records within a file; 40 – Serial files; 41 – Index-sequential; 42 – List processing; 43 – Inverted filing; 43 – Hash coding; 44 – Arranging files.

4. SEARCH STRATEGIES: RETRIEVAL 46

    46 – Enquiry formulation; 49 – Searching serially ordered files; 53 – Searching inverted files; 57 – Searching list-processed files; 61 – 'Natural language' searching; 64 – Precision and recall devices; 67 – Choice of strategy; 69 – Prompting systems (selection of descriptors); 71 – The rest of the system.

x   Contents

                                                                                             page

5. CONVERSION TO THE FINAL DESIGN, AND IMPLEMENTATION   75

75 – Integrating the parts; 78 – Flow-charts; 80 – Preliminary system agreement; 81 – Relationship between preliminary system agreement and final system; 81 – Manual work involved; 83 – Implementation, programming and testing; 88 – Understanding the computer.

Appendix 1. ELEMENTS OF PROGRAMMING, OR HOW TO MAKE THE COMPUTER DO WHAT YOU WANT IT TO DO   90

90 – What a program is; 90 – An imaginary programming language; 93 – Parameterizing; 95 – Machine-code programs; 96 – Programming languages; 98 – Ready-written programs; 99 – Multi-programming, real-time operation.

Appendix 2. GLOSSARY   101

Appendix 3. SPEEDS OF SOME PERIPHERALS   109

Reading List   110

Index   117

List of Figures

		page
1.	Suggested input specification form	21
2.	Sheet A of a two-part record description form	23
3.	A suggested form for recording the sources and uses of various record elements: the second part (sheet B) of a two-part record description form	24
4.	Showing where fields start (first method)	37
5.	Showing where fields start (second method)	38
6.	'Brother' and 'son' addresses in list processing	44
7.	Sequence of operations when searching a serial file using only 'and' logic	48
8.	Sequence of operations when searching a serial file using 'and' and 'or' logic	49
9.	Sequence of operations for item search using 'and', 'or' and 'not' logic, with planes shown numerically	51
10.	Inverted file search for A & (B or (C & D))	52–53
11.	A very complex enquiry handled by inverted-file system	55
12.	List processing: five records from the beginning of a list, giving for each only the first two descriptor fields	58
13.	Keyword 'page' dictionary	59
14.	Multi-processing with list structure	60
15.	Relative merits of three kinds of file design	68
16.	Four sub-records of one record of a descriptor showing lattice relationships	70
17.	Flow-chart showing that most programs are sort and/or sort/merge	73

xii *List of Figures*

		page
18.	Some flow-chart symbols used in input or output, and storage	77
19.	Symbols used to show jumps in flow-charts	78
20.	Arrow-heads used as jump symbols	79

Preface

This book is written for those who, knowing little about computer-based information systems, are faced with the necessity or desirability of making such a system for the storage, retrieval and dissemination of information. I assume that the reader is supported by the staff of the computer department for the programming and program testing, so neither of these is covered extensively. The book tries to enable the two sets of people – information specialists and system specialists – each to understand what the other is about; the book will also, I hope, make clear to the computer specialists why the information staff make so many and – to the former – so strange demands on the system.

It is also, alas, a regrettable truth that many information scientists think of the computer as a virtually all-powerful machine; they perhaps forget that it costs a lot of money and time to write a program and make it work, and almost as much to maintain it. They tend also to ask that the machine do a number of tasks (clerical or – even worse – intellectual) to which it is ill-suited; here, too, I have tried to clarify what should and should not be expected.

Much of the literature on mechanized information systems gives the impression that only intellectual problems of information organization are involved. This results from the fact that during the twenty or so years in which we have been making computer-based information storage and retrieval (ISR) systems we have been primarily concerned with the not inconsiderable intellectual processes in which our increased handling power has involved us. These will be dealt with in other books of this series.

To those who have already implemented their ISR systems and are now concerned only with making something better, this book will probably have little of value to offer, but to those faced for the first time with putting it on the computer it is my hope to say something of the practical problems they will encounter and to make their earliest explorations of the (almost) unknown land a little less frightening than they might otherwise be.

The bibliography attempts to give references to the theoretical foundations on which any information system will be based (and I hope that these theoretical problems will be kept to the forefront of the mind during system design), but this is primarily a book about the practical day-to-day problems of getting a new system working.

HELEN M. TOWNLEY

117 Wickham Chase, West Wickham, Kent

CHAPTER I

Preliminary Ideas

WHAT IS A SYSTEM?

If we are going to analyse a system we ought to be sure we know what a system is! The problem is, quite literally, one of definition, of drawing boundaries about the part to be distinguished, for it must be recognized at the outset that nearly everything is system. A gearing system is part of a bicycle; a bicycle with a man riding on it is a system; bicycles and cars moving along a road make up a system; a collection of roads and vehicles is a system.

In each case the system is a collection of entities interacting on one another. Each entity may be merely a thing, or may itself be a system and so one goes down the scale until one gets to the place at which there are no more systems, only things. And these things interact with each other in a recognizable pattern to make a system. The job in system analysis is:

to identify the object or objects of the system;

to identify the interactions by which the system works;

to identify the things involved in each interaction or, at a higher level, the sub-systems involved in each interaction.

For example, elements of a current awareness system are:

acquisition of items of information;

circulation of these items to individuals;

issue of bulletins about these items;

selective dissemination of information (SDI) about specific items to specific individuals or groups.

A change in any one of these can affect the functioning of some or all of the others; if the number of journals purchased goes beyond a given limit, the problem of circulating them changes, and it may become necessary to attach more importance to bulletin issue and/or the selective dissemination function. If an efficient SDI programme* is introduced it may be that the demand for current awareness bulletins drops to such an extent that they are no longer justifiable outputs of the system. Further, the nature of the total system is affected by the environment in which it operates: if the number of people served by the system extends beyond a critical point, or if their interests are widened, the system must adapt to these changes in the environment in the same way that a mechanical system may have to be changed if it is required to exert more force or to be lubricated by a different kind of oil.

A system, then, is an organization for the movement of materials (or information); its characteristics are dictated by the nature of the materials (or information) to be moved, the principles identifying what is to be moved whence, whither and when, and the environment in which the movement is to take place. Systems analysis† is the art of identifying all the elements involved in these three areas; system design is the art of defining the techniques by which the system shall be kept working. The problem in both areas is to ensure that the distinction has been sufficiently recognized between the system under examination and the environment in which it exists and of which it is itself a sub-system.

In all system design it must be remembered that the computer is not the 'giant brain' which it is sometimes imagined to be; it can only do what it has been programmed to do, and its proper utilization depends on the designers knowing what they want it to do and specifying the required work in great detail.

COMPONENTS OF DATA BASES

Let us look at ISR systems in another way. A data base or information bank (the terms are used as if synonymous) is a collection of entries. Each entry is about an entity, be it book or person, com-

* N.B. 'Programme' is used for the ordinary meaning of the word, e.g. 'a concert programme', 'a programme of exercises'. In English usage, 'program' is reserved for a computer program.

† It will be noticed that sometimes 'system' and sometimes 'systems' analysis is used. Where used at all by analysts, they are used indiscriminately; in fact, they usually just say 'SA'.

modity or statistic, and about its essential characteristics. At the very least it must contain for each entity:

a means of identifying it uniquely;

an organized description of it which may be used for searching.

It may also, but does not have to, contain:

an address of the entity: the place in which that entity can be found;

further, unorganized, description, not used for searching but provided for the searcher's benefit as an aid to his understanding of the retrieved entity, e.g. an abstract.

This question of organization is important: even the so-called 'natural language' systems perform some analysis of the input text and organize the words into concordances or thesauri for searching, though carrying the text itself within the system as 'unorganized' description.

The means of identifying the entity may be the same as the address of the entity; 'Nature, 262 (5564). 8th July, 1976. 84–85' is, as it happens, a succinct but complete identification (apart from author and title) of an entity and also of the address of the entity. But in large systems it is often found that the identification can be expressed in too many different ways for the system to handle and a simpler, often numeric or alpha-numeric, identification is allotted.

Descriptor/entity ratio

One of the factors that can be of great importance in defining the type of system to be used is the relationship between the *number of entities* and *the number of descriptive terms* used per entity. Compare, for example, two directories, *Kelly's Directory of merchants and manufacturers* and *Kompass (United Kingdom)*. Each is a tool for retrieving the names of companies according to the products they make or sell. *Kelly*'s has 1·3 products listed per company; *Kompass* has 23. *Kelly*'s can thus take the form of a simple list of products with, under each, the names of companies; *Kompass* is a complicated co-ordinate index. A telephone directory is an information-retrieval system, but its characteristics dictate the form it takes; these are (a) a very large number of entities, and (b) a very nearly one-for-one relationship between the number of entities (telephone number) and of descriptive terms (name-and-address). No other form is conceivable than a simple list with random access (which means that one can look anywhere with equal ease). A statistics student new to ISR once went to a lot of trouble to assign six or seven indexing terms to each of a set of lectures and then submitted the result to a clustering program,* only to find that each cluster contained terms applying to only one and the same lecture; in other words, a simple look-up table would make a perfect retrieval tool for that system. In brief, the shape of the retrieval system is defined principally by

the number of entities;

the number of descriptive terms in the descriptor language;

the ratio entities per term;

the ratio terms per entity.

THE ELEMENTS OF DOCUMENTARY ISR SYSTEMS

The documents and the information they contain

The 'materials or information to be moved' in an ISR system are normally documents, information about documents, information contained in documents, and information about the information in documents. (They may also index specialist skills, experience or sources of information.) The total system must pay attention to the physical documents – their acquisition, distribution and storage – as well as to the information contained in the documents and its relevance to the user. When people talk about information storage and retrieval (ISR) systems they usually mean information about (a) documents and (b) the information contained in the documents.

Information inherent in the document may include:

an author or authors;

a title;

a code and number (e.g. for patents and reports);

---

* 'Clustering' is a computer operation which scans the records of (in this case) documents and separates them into groups of similar records having a number of descriptors in common.

Preliminary Ideas 5

a publishing body or a source (and if necessary its location – e.g. volume and page numbers – within that source);

a date of publication;

a statement of author affiliation;

the country in which (a) the reported work was performed; (b) the author normally lives; (c) the author normally works; (d) the document was published;

the language in which the document is written;

the physical characteristics of the document (e.g. in fiche);

the number of useful pages;

text (words, phrases, sentences, paragraphs and perhaps headings at different levels);

figure and table captions; the contents of tables;

citations of previous works;

an abstract or summary;

an index.

Additional information which may need to be provided about the content of the document is:

the 'classification' (for security);

the name, place and date of any conference at which the paper was presented.

The decision as to which of these elements is to be included in the basic record of the ISR system is one which none but the information scientist can take, and then only after he is sure that he has comprehended the needs of the users for whom, ultimately, every ISR system is built. In some circumstances he may wish to include even more elements: for instance, the grant under which the work which is the subject of the document was funded. But it is essential that at an early stage in the analysis a list of all the elements be drawn up, so that an informed selection can be made, suited to the uses to which the material will be put.

THE FUNCTIONS OF AN ISR SYSTEM: WHAT IS TO BE MOVED, WHERE FROM AND TO

An information bank is rarely worth having if it is to be used only for retrospective searching when an enquiry is received.* So much use has been made of the phrase 'information retrieval' that we forget that the proper description is 'information *storage* and retrieval'. The whole object of an ISR system is to store information for retrieval as required; the amount of time and money spent on answering enquiries is minute compared with that spent getting the information, putting it into searchable form, and storing it. To get a worthwhile return from this labour, a system must be used for all it is worth.

There are four different functions, each making different demands on the system and affecting the system design:

Retrospective searching

This operates over the whole, or defined large parts, of the information base, searching a (proportionately) large number of records for a (proportionately) small number of enquiries. The number of retrieved items depends on the nature of the enquiry, of course, but in addition is governed by the selection of documents incorporated in the base, the way their 'organized' description has been formulated and the skill used in formulating the enquiry. Increasingly, in on-line systems, the computer is being used to help in enquiry formulation.

SDI (*selective dissemination of information*) *searching*

This operates only over a small fraction of the total information base – the recent acquisitions; and in proportion to the number of documents searched the number of searches made (i.e. the number of individuals whose 'interest profiles' are stored) is very high indeed. This sometimes dictates a different search strategy from the retrospective search: the latter being performed on the highly formatted information in its permanently stored form; the SDI search being, perhaps, performed simultaneously with the formatting.

---

* This is based on economic grounds: the 'intangibles' such as a felt want of some comprehensive, central collection on a given subject may quite outweigh the economics of the situation.

Group profiles for current awareness

These put to the newly-received documents very generic 'profiles' each for a group of users; the ratio profiles per document is a lot lower than that in SDI, coming closer to the retrospective-search proportions. The type of output required, too, is more like that of a retrospective-search system, being more detailed than that commonly required in a personalized SDI system, apart from there being many more references per profile than are produced by either SDI or retrospective searching.

It is not unusual in systems incorporating a bulletin-production program for class numbers as well as subject-term descriptors to be allocated to the documents. In these systems the broad class number may select the current awareness bulletin required and its sub-divisions may be used to arrange in a desired sequence the items in the resulting output, whilst the indexing terms are used for retrospective and SDI searches and, perhaps, for inclusion in the bulletin output as additional information, possibly in lieu of an abstract.

'Desk-top' retrieval aids

By 'desk-top' is meant the idea of retrieval aids multiplicated so as to be available on individuals' desks, not just in an information department.

In many ISR systems the production of printed indexes, feature cards, dual dictionaries (see Glossary [Appendix II]), and other desk-top tools was not merely a by-product of the establishment of the system, but an afterthought. Occasionally a system grew up designed first to produce such tools and only later was it modified to produce as by-products current awareness, retrospective-search, and SDI services. Index-production programs are not themselves designed for answering questions immediately; they produce question-answering tools, often with far-from-simple transformation algorithms (procedures) preceding the program which sorts the result into a simple output formulation.

Any, or all, or any combination of these functions can be served from one information base. At the time of first implementation of the new ISR system only one may be contemplated, but it is desirable to bear in mind from the start that one or more of these may be required and to build into the system at least a groundwork for producing them later – to leave a hook, as it were, on to which the new element can later be hung when required.

VARIETIES OF INFORMATION SYSTEM REQUIREMENTS

In the last section we noted four classes of use of an information storage and retrieval system, each making its own slightly different demand upon the system; and recognized the importance of identifying, early in the system analysis phase, what mix is required in a given case.

These classes of use are not, however, the only factors which can influence the shape of the eventual system. Others must be categorized also, and these fall roughly into the following classes:

the subject matter of the data base;
the kind of information required;
the viability of the information stored;
the expected growth-rate of the base.

The subject matter of the base

Although most ISR systems are concerned with the information stored in documents, this is not true of all information systems: quite massive information banks, for instance, comprise the identities of chemical compounds together with records of tests made of their properties, and their action on living organisms. Such a system uses a completely different kind of retrieval language from the documentary system and may require completely different system design. So may other systems which index skills and sources of oral information. Some of the principal classes are:

Conceptual information. The base is normally used for seeking information recorded in documents and the 'organized description' (see p. 3) is primarily concerned with concepts. A base designed to answer enquiries on information science might be expected to deal, for example, with what has been published in the last three years about the use of natural-text analysis to give a controlled thesaurus of indexing terms. The only finite element in this is 'three' – all else depends on the use of concepts defined more or less clearly in the minds both of those who feed information into the system and of those who seek to extract information.

In systems of this kind there may well be a considerable range of such descriptive terms attributable to any item; some papers may have only four or five descriptors or keywords whereas others may have up to a hundred or more descriptors. With these systems it is best not to provide for a maximum number, but to leave the records organized in such a way that any number per item can be accommodated. It is, however, still useful for the analyst to have some indication of the distribution of descriptor usage, e.g. average number of descriptors per document, standard deviation and total range. (This may also be done for types of document, e.g. patents.)

Information about objects or substances is often wholly factual and concerned with finite values – dimensions, voltages, density, mass, and so on – which may need no thesaurus of terms. Nor is there the wide range of number of terms as there is in conceptual systems. The type of data recorded for a class of objects or substances is much the same over the whole class and it may be possible to say, for instance, that for each object there will be seven values of temperature taken at each of seven pressure values; so it can be known beforehand that in this part of the record there will be forty-nine values – never less, never more – and that each will fall in a range that can also be pre-defined, at least as to the number of digits required to express it. If the objects concerned are human, then there is possibly only one descriptor that has two values only – alive or dead!

None the less, a degree, often quite considerable, of conceptual information can exist alongside the information expressed in quantitative terms. An information bank about organizations, for instance, in addition to the finite information on their financial matters, may need the title of the principal executive, which may be 'Managing Director' in some cases, or 'General Secretary' in others. Moreover matters of judgement may have been added to the record – that the company is diversifying, or declining, for instance – in which case one is again up against the problem that there may be a widely varying number of such descriptors attached to the items.

Briefly, then, the main distinction is between factual and

conceptual information banks or data bases. With the former, it is often easy to pre-define the number of descriptive elements that will be required, although a search program will probably have to be able first to locate the element and then to examine the value contained in that element. For a conceptual bank there is an almost unlimited range of descriptive terms which might be attached to any item, but the search program normally is required to discover only whether given terms are or are not present. Sometimes a value may be attached to a descriptive term, but this is exceptional. Many information banks do, however, comprise both classes of information. These must be examined carefully: it will generally be found that the two can be treated as two separate information banks. The reason for this distinction will become clearer in the discussion on file arrangement in Chapter 3 (pp. 31–45).

The nature of the response

Information systems can be classified by the sort of search tactics implied in the sort of response. There are four main groups:

Look-up systems. Even in something as simple as this, one can distinguish two kinds of query which make different demands on the same file. That which asks, 'What boils at 100°C at a pressure of 760 mm of mercury?' is probably simplest met by a file listing in a specified order (ascending value of temperature) the temperatures and, with each, the compounds for which that temperature is the boiling/melting point. On the other hand, 'What is the boiling point of water?' requires a file of compounds organized in a very different way in which 'water' (or H_2O, or however else you wish to express it) is easily found, and in which the 'water' record includes the boiling point. But in either case, the principle is that the file is ordered in such a way that one can go at once to a point, which is implicit in the enquiry, and at that point find the required information.

Information systems require a search through a number of records for those which match the specification in the enquiry; but once any such record has been found, the result is merely to output the required information. This information can range from merely the physical location of the answer to the enquiry ('You can find this in *Computer Bulletin*, 14(12), December 1970, at page 405'), to the actual answer in the form of a print-out of the exact words of the document. With the new technologies coming along, this may take the form of a facsimile (commonly abbreviated as fax) transmission of the content of a microfiche held within the computer complex.

Derived information systems. In these systems, once the matching records have been located, some computational work has to be done. For instance, having located records of compounds with a given specification which have given biological effects on the circulatory and/or the nervous system, output the numbers and percentage affecting circulatory only, nervous only, and both, and express all as a percentage of the total number of compounds recorded on the file. In this kind of system neither the search parameters nor the computation requirements impose any great demands on the programs, but the combination of the two makes this a unique class of information system.

Inferred information systems. These systems are as yet in their infancy; in them, the computer is not expected to have the answer stored in any specific records, but it is expected that there will be records holding perhaps only a small part of the total information, and that logical operations on these information fragments will permit the answer to be deduced. An unreal but not untypical example supposes that the following statements are held variously among the files:

Record 1: Company A is wholly owned by Company B;
Record 2: Company C is wholly owned by Company D;
Record 3: Company F owns 51 per cent of Company D;
Record 4: Company F has bought Company B.

A human brain has little difficulty in answering the question, 'What are the subsidiary companies controlled by company F?', but it will readily be seen that the computer program to undertake this kind of work requires both search facilities and logic facilities. This distinguishes it clearly from the other three classes mentioned above.

The viability of the data or information

Three quite distinct classes of information retrieval system can be distinguished under this heading. They can roughly be described as static, fluctuating and ephemeral.

Static systems are those in which the item record and the contents of that record are virtually unchanged for the whole of their life in the information base. Nearly all conceptual documentary information systems fall into this class. Exceptionally a part of the record may be varied: one might add, for instance, references to works which conflict with or confirm the work quoted in the document of which this is a record – a 'citation record'; or a record may be put on to the file before it has acquired all its descriptive terms, which may be added later. But to all intents and purposes these systems are completely unchanging. This, of course, requires the systems analyst to obtain some sort of judgement of the growth rate of the file: is it linear or exponential, and what is the maximum length of time a single record can be expected to remain on file? The information scientist must be prepared for these eminently understandable questions – although he seldom is!

Fluctuating systems. In these, the record stays on file for a long time, but its contents are likely to be altered from time to time, either by adding more information or by changing the information. Laboratory records are often of this type, being records, for example, of work on a series of individual items with variables in the conditions of testing, or additions to the test conditions. In these cases the information scientist must be able to tell the analyst what data are being recorded, at what rate a record may be expected to grow in length and how often a single data element can be expected to be changed during the life of the record.

Ephemeral systems are those in which both the data and the record itself are on the file only fleetingly. These have been nicknamed 'marriage-bureau systems' since commonly the record is deleted from the file once a match has been made, but systems such as airline-ticket reservation systems in which the record is removed by a given date also fall in the class.

Response times

As far as computer systems are concerned there are only two possible classes under this heading: 'now' and 'not now'! The former are the so-called 'on-line' systems (because the person putting the enquiry is operating over some form of direct link to the computer), the latter the 'batch' systems because in these the enquiries are saved up till the end of the day, week, fortnight or suitable period and then put together into a batch for computer processing.

It is not merely the immediacy of response required which is the sole criterion of whether or not an on-line system is to be undertaken (even if one leaves costs out of the equation); on-line systems offer much more than a quick answer. In particular they make it possible for the computer (or, to be more accurate, the systems designed for the computer, but I am not going to use this circumlocution every time) to operate in specified circumstances to prompt the enquirer to frame his enquiry or to help the indexer to attach the most exact word as an indexing term, and generally be of assistance in both compiling and using the information base.

Whatever may be the reasons for choosing one or the other system, the fact remains that the selection has a far-reaching effect upon the type of system design. A batch system, putting a number of enquiries to the information bank at the same time, might well be prepared to match them all first against record 1, then against record 2, then against record 3, and so on, but an on-line system must be prepared to handle any enquiry at any time. The design considerations based on how long it takes to access any one item within the computer memory are completely different, and this difference will affect both the way in which the files are designed and the way in which they are held in the computer memory.

Effects on design

All these variables affect the design of the system. It is, indeed, little short of miraculous that 'package' programs for information storage and retrieval are possible at all! They are available partly because of the 'modular' nature of many of the program sets which make up the package, so that a variety of systems can be built out of a number of modules, each designed with some special group of variables in mind. It must always be remembered, however, that an analogy between one's own program and a package, and a made-to-measure and an off-the-peg suit, is imperfect. The off-the-peg suit does not (unless you have chosen conspicuously too tight a garment) require you to change your shape; you just go on with an imperfect fit across the shoulders, as it were. But once you buy a package program, you have to fit it; and it cannot (easily) be altered by that little tailor round the corner, to fit you more exactly. You may wish desperately that you could in some cases have more than the sixty-eight descriptors which is its maximum per document, but the fact will remain that you cannot have sixty-nine. Package programs can save you large amounts of expense on design and programming, but they should never be chosen without a full examination of the nature of your own system and a full understanding of what you must have, what you would like to have, and what you definitely do not want. Expenditure of time, trouble and money on this preliminary work is essential.

CHAPTER 2

The Initial System Study

INTRODUCTION

The preceding chapter has been concerned with setting the scene in which a computer-based information storage and retrieval system is evolved. Now it is necessary to consider the form of that evolution. The several steps in producing a computerized system are:

initial survey and system study;
system agreement with the user;
detailed analysis of the problem;
system proposed to and agreed by the user (feasibility report);
system design;
programming;
program testing;
system testing;
documentation;
hand-over to user.

System designer and information staff

The system designer particularly needs to understand why the objectives of the system are what they are, and it is here more than in any other aspect that the information staff fail him. The initial survey and system study are crucial and are best made by a member of the information staff (an expert in hardware or software is not required at this stage, although an understanding of them is essential). He should be able to recognize the limitations of the particular configuration of equipment that is proposed, and be able to talk with the computer staff on their own terms, though with less expertise than they who will be making his system work.

Most information storage and retrieval systems differ from the majority of computer systems in that they are designed to serve a goal qualitatively and quantitatively different from that served by the existing system and its sub-systems. However, the basic questions posed during the survey stage of a proposed mechanization of an existing system serve as an excellent guide to the work performed in the initial stages of an ISR application.

Statement of goals. What are the goals of the system? What are our objectives in providing a system of this kind? How do they rank in order of importance to the organization? How do they fit into the system which is the organization?

Statement of requirements. What are the requirements which must be met to satisfy these goals? Who, specifically, is to benefit by the provision of what, when and in what circumstances?

System study: inputs. What data will provide the means of satisfying these requirements? Whence and when are they available? What is their nature and obsolescence? What is their relationship to each other?

System study: outputs. What outputs from the system will satisfy the requirements? How are they obtained, deduced, calculated or assumed from the originally available data? What is their nature, how often are they required and in what circumstances are they required? What records need to be kept to provide the outputs?

Is a computer system really necessary?

Throughout, the need for a computerized system should be questioned. It may well transpire, during or at the end of the initial system study, that some form of manual system would be more appropriate; but painstaking attempts to answer the foregoing questions will help to elucidate the position and make a better-based judgement possible.

STATEMENT OF GOALS

The statement of goals needs to specify:

What is being done at the moment, and why. It should deal

The Initial System Study

individually with the several elements of the information service, but only in general terms.

In what way the existing service fails to meet its own goals, and what reasons can be attributed for this.

In what ways the goals of the existing service are insufficient for the organization. Were any of them recognized as being desirable during the evolution of the existing system, but specifically disregarded and, if so, why? What goals now considered have never been considered previously? For what reason are they now being brought into question and what are their relative values in relation to the goals of the organization?

What goals are desirable but considered impossible of achievement either because the techniques are not at present available, or because it is believed that the cost of implementing them will be too heavy in terms of money and/or manpower.

The statement should end with the constraints to be put upon the proposed system in terms of the financial support which can be afforded and the equipment which can be utilized; if possible this should relate to at least a five-year period ahead.

In all goals suggested for the present study, an indication should be given of their relative priorities.

It should be noted that this statement does not specify the use of any specific equipment for a specific task: it merely sets limits by which one obtains a guide to the types of equipment which need to be considered. The indication of the apparently unachievable goals is desirable, partly to emphasize the underlying philosophy of the service as a whole, but partly because the later work might show that some are in fact possible and might indicate where 'pegs' should be left by which, even if not immediately to be implemented, they may in future be attached to the system to be designed.

The statement of goals, therefore, is not compiled merely by the head of the information department out of his own experience; into it he must bring his top management, his computer staff, his users and his own information staff. It is to be the most important single document in his hands for obtaining (and keeping!) backing for the project and for keeping control of the evolving system as it comes into being. It must be wholly authoritative and regarded as being as much binding on himself as on the other sections of the organization who will be using it.

It may be argued that this is a counsel of perfection, and perhaps it is. But too often a project sails off hopefully into the sea of mechanization and runs aground on the sands of differing opinions about the desirability of doing this or that, or is nearly wrecked by the realization that management will not stand for some particular aspect which has only just been recognized as being feasible but which would be *so* valuable. If it is to be valuable, it should have been thought of from the start, even though later to be written out of the feasibility study on grounds, say, of expense. But once the project has got under way, there ought to be no dispute about what it is trying to do.

STATEMENT OF REQUIREMENTS

The preparation of the statement of goals was a matter for the policy-makers; the preparation of the statement of requirements is where the system analyst begins his work. The statement of requirements must specify the demands which will be made on the system and the system components that are required to satisfy them.

Demands may be made on the system by people requiring information and reports, or by systems which include people, e.g. management. For every demand it is necessary to specify its nature and the circumstances in which it arises.

It has been said of ISR systems that they collect information with a view to answering unanticipated questions at unanticipated times. If this were the case, there would be no such thing as an ISR *system*, but the statement does contain an element of truth which must be recognized. The closer the analyst can get to precision the better, even though complete precision may be out of the question. He can only estimate the degree of flexibility, for instance, which can be built into the indexing language and into the forms in which enquiries can be put to the system, but he cannot avoid the necessity to give as good an indication as he can. He cannot probably be exact about the form in which retrospective enquiries will be put to the stored records of documents, but he must give, as best he can, an indication of the nature of the matching process and the enquiry formulation to be used. He cannot say precisely what will be the result of putting an enquiry

The Initial System Study

to the stored records, but he must be prepared to specify the kind of information to be given and the form in which it is to be given, and to give an estimate of the volume of such information.

We might pause at this point to consider how far the analyst need go in this initial system study. The statement of the goals of the system will have covered such matters as whether an SDI facility should be provided; the feasibility study will be making recommendations on what kind of file structure is to be used for retrospective searching. The analyst's function is to bridge the gap between these two, to elicit from the statement of goals a complete specification of the things that must be done: the goals to be met; not, be it noted, how they are to be done, only what they are. 'How?' is determined later.

This stage of the work is best tackled by creating a hierarchical listing of the requirements. From the statement of goals, perhaps six major requirements can be identified. For each, it is necessary to identify *its* major requirements, and so on, not *ad infinitum*, but until the analyst cannot identify a component which cannot be described in terms of:

its nature;

the circumstances in which and frequency with which it arises;

the demand which it satisfies;

the frequency with which that demand arises;

the records which must be maintained in order to fulfil it;

the operations which must be made upon those records in order to fulfil it;

the decisions which have to be made during the course of these operations;

the reports which are generated as a result of these operations.

As far as the last is concerned, it should be noted that a report being generated indicates a demand being met; each such report, therefore, will require the pursuit of its own statement of requirements.

During the course of this part of the study, the analyst will be building up lists of inputs and outputs and records which must be maintained. By the time he has finished, he should know what records and reports are needed, what operations are required to generate them, what information they contain, whence the information they contain is derived, and why and how the contained information is used.

SYSTEM STUDY

Records made during the system study

In the feasibility study a much deeper examination of the nature of the following elements will be required, but now is not the time. At present we need only generalities and the simple forms suggested in this chapter should suffice. Here an *input* is information or data periodically available to the system but not generated within the system; an *output* is information or data generated by the system and communicated to people or systems outside the system; a *record* is information within the system held in identifiable form during the process of transformation of input information to output information; or information continuously held in the same form within the system for use during processing.

This distinguishes between the variables upon which the system operates – normally document records in our case – and their transforms; and the reasonably constant elements which the system stores and uses, such as thesauri and dictionaries.

This distinction between 'input' and 'record' must not be taken too literally: inputs (and for that matter, outputs) *are* records. There may be occasions when an input goes into the system and comes out again at the other end virtually unchanged, perhaps only with the addition of a code or class number. But for the purposes of getting a written picture of a system, I have found this distinction invaluable.

One last general comment: never make a note of anything without dating it. This applies to written recommendations and to notes made during the course of a discussion. During the course of study, many pages may well be written about changing forms of one record: to have each dated will make the study much easier!

Inputs (see Figure 1)

Name of input. Give a descriptive name to the input; if it is commonly known to the users of operators of the existing system by some non-descriptive name, give that in brackets and use both in

NAME

SOURCE

FORM   printed/typed/p.card/p.tape/m.tape/fiche/film/other (specify)

VOLUME   per

AVAILABILITY

ORIGINAL DATA ELEMENTS	USED FOR		
ADDED DATA ELEMENTS	USED FOR	ENTERED FROM	BY WHOM

1. Suggested input specification form

the text, e.g. Patent record (Derwent card). Both should be used because your system study is going to be read by computer staff and management staff as well as the information staff; the first would not recognize the common name and the last might be puzzled by the formal 'Patent record'.

Form. Only a brief description is necessary: 'punched card'; 'magnetic tape'; 'abstract periodical', for instance.

Availability. State when records of this nature will first become available to the system (if they are not already). Give also the frequency with which the records are available. (This is usually only for items which are a complete file in themselves, e.g. a magnetic tape which arrives weekly. For the normal documents, whose frequency varies from daily to annually to rarely, no entry is needed here.)

Volume. Indicate the number of records per day/week/month.

Data elements. There is no need to be excessively specific here. A journal paper could well, for instance, be given just 'citation data' under the data element heading; it is important, however, to show every record into which data elements are entered which first came into the system via this input. Thus, if the citation data travel together right through the system, there is no need to specify any one of them individually; but if, for instance, title alone were to be entered into some other record (as it would if a KWIC were being made) then it would be necessary to show it separately in the input specification, which would then, probably, look thus:

DATA ELEMENT	USED FOR
citation data	document record
title (part of citation data)	KWIC record

In some cases, data elements are added to original material before it is input. Filing codes, for instance, are often entered on to press cuttings. Here it is necessary to show by whom they were added and whence they were drawn (if not from the imagination of the enterer).

The Initial System Study

FIELD Sub-field Sub-field	FREQUENCY in file or field Max. | Min. | Avge.			NO. OF CHARACTERS Max. | Min. | Avge.			TYPE OF CHARACTERS (PICTURE)	COMMENT
DOCUMENT No. year separator serial number	— — — —	— — — —	1 1 1 1	7 — — 4	4 — — 1	6 2 1 2	NN/NNNN NN / NNNN	Last two digits of year of receipt of document; serial no. within year from 0001 – 9999
CLASS NUMBER	2	1	1	—	—	7	NNN.NNN	
TITLE	—	—	1	135	16	108	Alphabetical, including spaces, punctuation	
AUTHOR author name annotation institution	3 — — 1	— — — —	1.25 1 0.1 0.5	 20 10 50	 8 2 15	 12 4 20	 Alphabetical Alphabetical	Includes 'ed' 'comp' etc
CORPORATE AUTHOR	1	—	0.01	50	15	20	Alpha-numeric	Cannot co-occur with AUTHOR
SOURCE	1	—	0.999	200	95	130	Alpha-numeric	
KEYWORDS keyword	35	1	7	 50	 2	 37	Alpha-numeric Alpha-numeric	
SIGNAL	1	—	0.001	—	—	1	1 or zero	Indicates that keywords not title are to be KWICed

2. Sheet A of a two-part record description form

Records

Information is needed about the compositions of each record type (there may be several types to one file) and the use to which each field may be put (Figures 2 and 3). Several of the figures given in Figure 2 can only be guesswork – indications of frequency rather than detailed counts – but they should give as accurate as possible a representation of the state of affairs. This form gives a picture of the composition of a class of record and of the fields and sub-fields it comprises. It is not necessary to specify a sub-field if that separate element of a field is never going to be used alone. For instance, in the record shown in Figure 2, the year is shown separately because it is used only in retrieval: in the KWIC listing, the numbers run from 0001 each year and the 'year' part of the code is not needed.

Sheet B (Figure 3) forms are made only as more information comes in: there is no need to enter a field name until it is known that that element is going to be used; once that is known, the record or output on which it is to be used should be entered

('What?' on the form) and the purpose and frequency of this use ('Why/When?'). It may be possible at this time to enter also the source (input or another record or – as, for example, the date – computer-generated) but often this has to be entered later. Fields can be used for all sorts of purposes (for instance, a class number can be used for sequencing a KWIC print and also as a search descriptor), and there is provision on the form for four uses.

Note that it is only when a field is used *from this record* that there is an entry for it in the 'Used for' columns: a single field may well, within the whole system, be used over and over again in a variety of circumstances. But the object of this form is to show what fields require to be kept together in a record.

The investigator is warned against creating too many or too few of these (at present) hypothetical records. The temptation is, at first, to make every one a single-field record, with a lesser temptation to make only one or two records for the whole of the material to be stored and used during the processing. Try to think of this as you would if you were creating a card index; let a record be a bunch of fields that naturally appear to 'go together' and could be expected to travel around together.

This may sound a very subjective way of setting about what is supposed to be an analytical process; perhaps this is as good a

RECORD DESCRIPTION, PROJECT_____ Page no.									
								Part B	
							Source and Use		
NAME OF RECORD									
Field name	Source		Used For						
		What?	Why/Who?	What?	Why/When?	What?	Why/When?	What?	Why/When?

3. A suggested form for recording the sources and uses of various record elements: the second part (sheet B) of a two-part record description form

The Initial System Study

place as any to indicate that system analysis is partly technique and partly flair. You work yourself into a job, getting a 'feel' for it, and your mind (as you are talking about the processes to be undertaken) starts building up a picture of the sorts of records that will be required. In the early stages of the study, you will probably create and destroy dozens of proposed record specifications, but they will gradually crystallize out, as it were, into a reasonable set.

Outputs

The data collected when recording outputs is very similar to that used for records and the same forms may be used; it may seem odd to provide a space for 'used for' in the end-product of a system. The reasons are partly practical and partly philosophical.

The philosophical reason is that it is essential to know every element which is required in an output before one can assess the relative importance of all the items of information that are to be communicated. Like politics, designing an information system is the art of the possible, and compromise with the ideal is going to be inevitable. One needs, therefore, to be armed with the fullest information about the justification for each particular demand.

The practical reasons are: (a) that despite taking every care in trying to identify the several sub-systems from the statement of goals, there can arise during the study a recognition that another sub-system is required, and this arises from the examination of the 'Whys?' of the outputs; (b) to assist in the differentiation of the short-term and the long-term goals: these will have been broadly defined in the statement of goals, but in the design stage it is going to be necessary to design the immediate application so that the more remote applications can be brought in without disrupting the (by then) existing system. Indications of this nature in the initial study are therefore invaluable.

Approaching the system study

The statement of goals must come first. However, if the impression has been given that the four elements (statement of requirement, specification of inputs, specification of outputs and specification of records) are sequential, please correct it now. Apart from the initial study of goals, they are simultaneous. And since it is greatly to be desired that the analyst should be present during all discussions about the formulation of the statement of goals, it can be taken for granted that he is able to begin the system study as soon as the goals have been defined. Indeed, both he and the system designer will have had an impact on the goal definition, since that definition is expected to include statements about what is presently considered feasible, and these will be based on the contributions of, among other people, the analyst and the designer.

Analyst and designer (who may be the same person) have therefore begun their co-operation before the former ever gets down to the details of his system study. At present each is more concerned with learning what the other has to communicate than with learning what he is communicating; the analyst with little or no computer experience should make a point of regular if informal discussions with the system designer (who may also be the programmer) on the way he sees problems and their (information-oriented) solutions.

The greater part of the analyst's job is to make other people talk. For information on the inputs to the system he must go to the people who are handling the inputs at present, watch them at work, ask them to describe what they are doing and why they are doing it, encourage them to expound on the difficulties and on the things they would do differently if they had the chance. Where the inputs are not yet being handled (as, for instance, where it is proposed that a tape service is to be bought) he should familiarize himself with the way in which the data are prepared and the form in which they will reach the system. (He must ask the awkward questions about the reliability of a tape service, but he will find the suppliers are normally prepared to answer honestly; it is to their advantage that he design a system which handles their outputs efficiently, and provided he uses that tact which is an essential of his job, he will obtain their co-operation.)

It is important to talk to all levels of workers, and to more than one at each level. Try to recognize the difference between a fixed idea and the suggestions of a mind open to question: this can best be done by comparing the approach of different workers to the identification of their tasks and of the records they handle.

Following complete comprehension of the statement of goals comes the differentiation of the major sub-systems: the approach to the interviewing should be systematic, following through first one, then another, sub-system. This could mean several visits to the same people, to ask different sets of questions; and this can induce the hostility which is the enemy of good analysis. Plan each interview ahead, therefore; identify the various sub-systems about which you will wish to be talked to and have the main sequence which you wish the conversation to follow clear in your mind before you start. The fact that you are going to be wandering around asking questions should have been made quite clear to all staff from the start, and they should have been told why, but even so, preface your first meeting with anybody you interview with a statement of why you are making the survey, and why you have come to him in particular, showing the place you think his contribution will take in the total picture. There are two dangers that arise from antagonizing the people you are questioning. One is that they will not give you the information you seek. That is important enough, certainly; even more important is that here, right at the beginning, you may induce a hostility not merely to you, but to the project, which can become serious when the time comes for the implementation of the system. Never talk down, even to the most junior office-boy; your task is to find out about his job, and since he has done it and you have not, he knows more about it than you do even though he has never tried to put it into words and has perhaps never visualized his part in the system.

The amount of notes taken during an interview varies from analyst to analyst. I prefer to use a spiral-bound notebook rather than separate pages, so that I can keep my rough notes in the sequence in which I made them. During the interview, I make, correct, discard and re-make a number of versions of each input or report specification, numbering each 'issue', so that when going through my notes I can trace through the various drafts I made of the form. Ideally, one should write up each interview as soon as it is completed, but this is seldom possible. I do, however, try to make an as-good-as-possible specification form of the inputs and records discussed during the interview as soon as I can and indicate on the back the principal difficulties encountered and suggestions made by the person interviewed. To each input specification I attach annotated specimens of any forms etc., which he completes. These usually suffice and I rarely need to write up a formal report of any interview.

Outputs are a different matter. However specific the statement of goals may have been, the wildest of ideas will be produced when you come to ask what should be the outputs (except for those required for the more formal parts of the organization, the accounts, stores and similar departments). Many of the people you are talking to know that computers can store graphics, so why not output the results of the answers to some of the questions in graphic form? Some systems store full text, so why should not we? Would it not be better to have a remote microfilm store in every laboratory, computer-accessed to produce the frames holding the answers to the on-line questions?

Here the analyst is presented with a different set of problems: he has moved over from passive enquirer to influential discusser. He has the statement of goals to support him in restraining the wilder flights of fancy, but his own authoritative statements of what is or is not feasible in the context laid down by the goals will be more significant to those with whom he is discussing the outputs. Often a group discussion is the way by which he can best illuminate the question, 'How best is the information to be given out?'

GROUP DISCUSSIONS

Such a session needs careful preparation: its purpose is to elicit discussion of all the ideas gleaned previously and to validate them. Who takes part, what preparatory work is necessary; what confines are to be set to the subject under discussion; what the analyst suggests, and what the meeting makes of his ideas – all these questions need careful consideration.

Who takes part, and preparatory work

All the main providers of information to be operated on in the proposed sub-system must be present. Note that this includes not only the intellectual providers but the clerical workers involved. Obviously this does not (normally) go back to the originators of, for example, the documents under discussion, or all writers on the subject would have to be invited; but it must include all the abstractors and indexers (or a representative group, if their numbers or location make a full meeting impracticable) and the heads of their departments. There may be times when the actual originators of the information should be present: e.g. when the system is to

The Initial System Study

handle the reports which they originate. In this case, too, it may not be possible for them all to be present. The clerical workers involved are important: they can comment in particular on such matters as the ease with which the (proposed) forms can be read, grasped, copied, etc. and very often notice items of information which have been recorded from time immemorial without having been used for many years! All these points may well have been made during the original investigation, but to give them an airing in front of the group as a whole may be very productive.

It is important to ensure that some representative of the users of the resulting outputs be present to comment on the utility of the information which it is proposed to give them, and to expand on their reasons for wanting something they do want, and on the uses they will be making of it.

There should be at least one person representing the users of each of the proposed other sub-systems, to comment on the interaction of the sub-system under discussion with their own spheres of interest.

All this means that in planning the session the analyst/designer must have prepared and have available copies of all the forms (inputs, records and outputs) listed previously; and on his own copies at least must have notes on whom to encourage to participate in the discussion.

The confines of the discussion

These, too, must have been decided in advance by designer and analyst and should from the start be made clear to the participants. Where the discussion ranges outside them, it should be nipped in the bud, though it must be noted for future sessions or when analysing past sessions. Normally a 'brainstorming' session of this kind (in which people are encouraged to bring out the wildest ideas to stimulate discussion) will confine itself to one sub-system or – but only occasionally when the same parties are participants – to two or perhaps even three sub-systems. Discussion is kept to the point better if it is not allowed to be too broad in coverage, especially because a meeting of this nature is expected to spread widely and, to some (controlled) extent, randomly over the allotted subject, and a formal agenda is better avoided.

What the analyst suggests

His ideas must be presented to the meeting, but only as suggestions, not as firm rulings. He should show:

(a) the part this sub-system plays in the whole; e.g. what he believes it will receive, whence and in what form; what it is to be used for, and whither it is to be transmitted, and why;
(b) a more detailed description of the proposed inputs;
(c) the processes to be carried out on each element in transforming it into file and output format;
(d) why each element is thus treated/transformed.

The first job of the meeting must be to consider (d) above and this consideration must be exhaustive. The question, 'How often and why must each transformation cover what?' should be completely clarified (and note that this also involves a clarification of (b)).

When this is done (a) and (b) should fall into place permitting a more complete definition of what is to be handled: its nature, how often it is received and transformed, whence and how it is derived.

Discussion of (c) involves still further clarification of (a) and (d).

General

In this discussion the analyst and the system designer have one function only, having introduced their gleanings thus far: they are to act as idea stimulators and as flatterers! Their function is not to create ideas but to elicit thought on what should be done and when; nor is it their function at this stage to indicate the practicality of these ideas; in particular they must avoid pouring cold water on any that appear. The object of the whole exercise is solely to stimulate imaginative discussion of what is wanted.

On conclusion of the meeting, analyst and designer have a great deal of work to do in assessing first the validity of the ideas promoted and second the means by which the valid ones can be implemented. This should be completed in respect of one sub-system before going on to the next.

It should be noted that such a session is not essential, though often it is very productive. If it is not held, the analyst/designer has (have) to go through the same rather brutal examination of all that has been gleaned in the preliminary survey.

CHAPTER 3

Storage: Systems, Files, Records and Fields

ACCESS

Begging the question of what the system is to do, the problem of system design boils down to one of access. Information (data) is (are) stored somehow, somewhere, solely in order that it (they) can be *got at, used* and, usually, *put back*. This whole elaborate discipline is devoted to no more than ensuring that this is done efficiently, by which I mean that it does what needs to be done, in a mixture of time and cost that satisfies the user.

All information systems comprise elements of data. Like elements are identified and organized into

fields, which are organized into
  records, which are organized into
    files, which are organized into
      sub-systems, which are organized into
        systems.

And this is an over-simplification!

FIELDS

A date is an element which may be a single field, but may be three fields – day, month, year – needing sometimes to be accessed and used as three fields, and sometimes to be accessed and used as a single field. A record may be broken down into a number of distinct collections of fields which are in effect themselves sub-records. As was suggested in Chapter 1, in the ultimate analysis there is only one system that (possibly) is not part of a larger system, and that is the universe!

A field is a set of one or more characters with a defined meaning.

The nature of its content, though not the actual content, is predefined. A field might be specified as the field 'year' and given four digits. The nature of the field is thus defined, but the content as thus stated can range from −9999 to +9999. Often it is possible to set limits to the value of the field; that, for instance, it contains values from 1900 to 2000.

In addition to its name and its value, before it can be *got at*, *used* and *put back*, it is necessary to know its location in any record, and the problems here can be considerable. But these fall within the area of record construction and are dealt with later in this chapter.

It is also necessary to know the function of the field. In the early stages of system analysis fields' functions are related only to the function of the system. 'Date' is used for helping to identify a publication, for instance. There may be two important 'date' fields, however: one relating to the objective of the system; and another (for instance, the date of entry of the item into the library, used for weeding) which is not directed to the main objective of the system, but merely to its efficient functioning. Later, further fields may be added, related only to the internal function of the system, such as a field which gives the address of the first 'date' field. ('Address' is computer jargon for the location at which a piece of information is stored; 'to address' is to get at that piece of information.)

During system analysis and subsequent design, one is constantly adding to the list of fields to make up the (initially tentative) record, and in a later chapter some of the techniques for handling this decidedly messy collection of data are considered.

The rules that apply to a field apply, of course, to a sub-field. Taking again the simple case of a date comprising day, month and year, one may perhaps need to identify for each element:

the name:	day	month	year
the value:	00–31	00–12	00–99
the location, e.g.	characters 5 and 6	characters 3 and 4	characters 1 and 2
the function:	sequencing in SDI print-out	sequencing in KWIC print-out	identification during searching

The zeros in the 'value' field carry a warning; as far as day and month are concerned, they indicate that the field may be missing, whereas for years, there is good reason for the field sometimes containing oo. The system designer must decide what are to be his sequencing rules if in fact both day and/or month or day only are missing in any individual case.

The function and the location are inter-linked. If, over all, the records containing this 'date' field are to be used most frequently for identification of a period during searching, it is common sense to put the 'year' element at the beginning of the field. If, however, the file were to be subject to many chronological sequencing operations, common sense puts the reverse sequence, with day followed by month followed by year.

These particular examples are, of course, simple. But they represent a major part of the task of system analysis and design. What are the smallest elements of data which have to be dealt with? How are they used, and with what other elements are they used? How often do they need to be accessed and for what reason do they need to be accessed? It is only when these questions have been exhaustively answered that it is possible to proceed to the next step, the construction of the record. (Often only the information scientist can provide the answers; my attention was recently drawn to a case in which the information scientist had failed to supply the information that with Russian authors, the first character of a forename can be two characters when transliterated, not the one to which we are accustomed. Such information is vital to the designers.)

I would not have it thought, however, that no consideration is given to records until all the fields have been identified and defined. One's first approach to a job normally leaves one with a mental pattern of a set of records, labelled perhaps 'documents', 'users', 'index', 'SDI print-out', etc., for this is how one approaches a system in the analysis phase. In the design phase it is absolutely essential to have identified the fields fully, by name, value and function, and the position of sub-fields, before approaching the construction of the record.

RECORDS

A record is a set of fields descriptive of a single entity – though that entity may be a collective noun – there may be a set of records of type X describing respectively sets of records of types A, B, C, etc.

In compiling fields into records, a basic syntax or set of rules is necessary. One can only know that the string of characters 1, 3, 5 represents the number one hundred and thirty-five if one knows the basic rule that in the denary system digits to the left represent successively higher powers of ten. Similarly, a syntax for each set of records is essential to prevent their being merely a string of characters. There may be many kinds of record in a file, but for each there must be a distinct structure, completely defined. It identifies all the fields in the records and their relationship to one another: decides what kind of characters are to be found in each field, how long the field is, how many times such a field occurs in the record, and how that field is located within the records.

An important thing about a record is that it can be (and frequently is) transformed into another kind of record. The full text of an abstract is a record; a concordance is a transformation of that record into a set of records which have a completely different structure. The extent to which reformatting (which is computer jargon for transforming) one record into another is feasible depends on the syntax of the records. Simple record structures permit only simple transformations. Complex transformations or complex use demand either complex record structure or complex programming, or duplication of the same information in a number of differently formatted records in a number of different files.

The MARC record format (Machine Readable Catalog(u)ing, a tape format now widely used), designed to permit all men to do all manner of things with the record, is highly complex. The record structure of a feature (optical coincidence) card is very simple, but there is not a lot you can do with it except retrieve reference numbers.

Arranging fields within a record

A field can be *fixed* or *variable* in length. 'Year' could be a fixed field of two (73) or four (1973) or perhaps six (1973 AD) characters. But if it has been defined as a fixed field it must have that many digits in every record – you cannot have 73 in record A and 1973 in record B, if A and B are records of the same kind in the same file. But you could have 'year' as a two-digit field in the record on tape and a four-digit field in the record on print-out; the use of the record dictates the structure of its fields.

A variable field, as its name implies, is of variable length. The title of a document is normally put into a variable field, since one can never predict the number of characters in a title. It would be possible to say that there will never be more than a hundred characters in a field and for that reason make it a hundred characters long; but this could be very wasteful of space if the average number of characters is only fifty-five. A more economic approach would be to have a completely variable field: four characters for a document entitled ADAM and thirty-one for THE IMPORTANCE OF BEING EARNEST (note that the spaces count as characters, too: it requires a complicated set of programs and a dictionary for the computer to recognize an OF in the string . . . nceofbeing . . .). An alternative is to state that since, say, seventy-five per cent of the titles have less than thirty-five characters, a fixed thirty-five character field would be allowed, with a thirty-sixth character set as an indicator whether further characters were to be found in the adjacent field; the adjacent field would then be either a further thirty-five characters of title (which would complete all but very verbose titles) or the next field of the record.

Fixed fields. If a record comprises only fixed fields, it is easy for the program to address any one field, because the programmer knows just how far that field is from the beginning of the record. Similarly, if the record comprises only a fixed number of fixed fields, there is no problem in addressing the next record, since in every case it will be exactly the same number of characters along from the start of the previous record. But if the records contain both then one can never know automatically the start address of any one field or any one record, merely from knowing the start address of the first field of the first record.

A distinction must also be drawn between variant and invariant fields. If, for example, we are concerned only with this century, then the digits '19' in the first two positions of the 'year' field are invariant. Location within the record can be used to identify the invariant parts of a variant field. Thus an agricultural bureau used UDC numbers for descriptors and found that the majority of descriptors fell in one or more of the classes 51x 53x 57x 631x 632x 633xxx and that there was never need for more than one descriptor from each of these classes. So five one-digit fields were set up to represent the first five of these classes, and a three-digit field to represent class 633; the location represented the invariant part, and the value in the field represented the last digit(s) of the class number. A similar effect is found in the fields of a feature card: the location of the field identifies the field and the binary value (hole or no hole) is all that need be recorded in the field. Use of location to identify the field and its value is commonly used in computer applications. For instance, each 'bit' in a descriptor record in an inverted file can represent a document number, the value being set at one or zero according to whether or not that document is described by that descriptor. (A 'bit', a colloquial abbreviation of 'binary digit', is a digit space in binary arithmetic, i.e. the opportunity to say 'one' or 'zero', no other digits being used.)

Repeating fields are commonly encountered. A serial file of document records shows, in each record, a number of descriptor fields. If these are of fixed length (for instance, if four-digit codes represent the descriptors) then, knowing the address of the first, it is easy to access subsequent ones – each starts four digit positions along from the preceding descriptor field. Again, this presents little problem if, in every record, there are always eight such fields; but if there is a variable number, if in some cases there are only three, and in others twelve, some device must be introduced so that the computer does not treat the fourth or thirteenth set of four characters as if they were another descriptor.

One technique for overcoming this is to add a signal to each field, turning it into a five-character field; the computer can then look for the presence of this signal in each descriptor as it is examined. If the signal is set, the computer can go on to treat the next four digits as a descriptor; but if it is not, then the last descriptor is known to have been reached.

Another solution is to make a separate file for each class of repeating field and carry in the main record only (a) a note of whether or not repeating fields for this record are carried in their own file and (b) the address of the first. This means that you can have a fixed field in the main record and a variable number of fixed fields in the repeating field record.

Repeating fields of variable length present a slightly more difficult problem. Preceding the actual data in one field, one can put a fixed sub-field which has two elements, one being a signal indicating the nature of the field and the other giving the start address of the next field.

The idea of 'tagging' a field with its length can be carried a stage further: a series of fixed-length fields can be put at the beginning of each record, each (since they are fixed length) in a known location. In each of these fields can be given the start address of the data fields. In the example in Figure 4 the record

Characters	Field*
0–4	Record number
5–7	Start address of author(s) field (127)
8–10	Start address of descriptor(s) field (170)
....	
.....	
127–169	Author(s) field
170–...	Descriptor(s) field

* examples given in brackets

4. Showing where fields start (first method)

starts with a fixed field for the record number, then two fields, the first giving the start address of the first author field (at character 127) and the second the start address of the first descriptor field (at character 170).

What this method does not do is indicate how many author fields and how many descriptor fields are present. There might well be three author names in the space between 127 and 169.

To get over this, an index can be put at the beginning of each group of repeating fields. Like the index in the leader fields at the beginning of the record, this, too, can comprise fixed-length fields; there need to be as many such fixed-index fields as there are data fields, plus one, as in Figure 5.

To summarize, then: there is a fixed-length field at the beginning of every record giving the start address for every class of field; and there are a series of fixed fields at the beginning of the string of data fields in each class, which give the type of field (for extra security), the number of fields and the start addresses of fields, in that class. This is the basic principle of the MARC format, used not only in MARC but very widely where tapes are interchanged between organizations. It is a complex structure, but its very

Character positions	Contents	Meaning
127–129	A04	A – this is a set of author fields 04 – and there are four of them
130–132	142	first author name starts at 142
133–135	151	second author name starts at 151
136–138	159	third author name starts at 159
139–141	165	fourth author name starts at 165
142–150	VALLEE,J.	⎫
151–158	JONES,G.	⎬ the author names
159–164	GEE,A.	⎪
165–174	DEUTSCH,A.	⎭

5. Showing where fields start (second method)

complexity makes it extremely flexible. It makes for rather long records, but they are nothing like as long as would be necessary if every field had to be as long as the longest string of characters it were likely to contain, and there had to be space left, in the case of every repeating field, for the maximum number of fields there were likely to be of that class. Additionally, if at any time there appears a record which needs more than the maximum allowed for (if, for example, it requires space for thirty-five descriptors when the system design only specified a maximum of thirty) this new record can be accommodated without any trouble.

There is the additional advantage that if all exchangeable tapes were made up to the same format, the same program could handle them. (But this is not completely true: for instance an ICL machine receiving tapes prepared from an NCR computer would have to be programmed to translate the character encoding of the tapes into the character encoding used by ICL.)

The factors affecting the structure of a record are therefore:

the nature of the record in terms of the types of fields it comprises;

considerations of the space available (a) to hold the file in backing store and (b) to hold the records in core during manipulation, and the cost of that space;

the manipulations and transformations to be made; and hence;

the importance of ready access to any specified field.

Storage: Systems, Files, Records and Fields

Is it necessary to repeat that these last two depend on the needs of the users revealed during the initial investigation? However, these cannot be the sole considerations: some sense of priority must be exercised, or the system designer will find himself creating a system which goes to great lengths, wasting computer storage space, running time, input time, record-creation time, to perform a function required once a year by one user out of the hundreds for whom the operation is conceived.

FILES

A file is a collection of records, grouped together for convenience in accessing the data elements which they comprise. Types of file can be:

document files (various kinds of document may need various kinds of file);

user profiles;

vocabulary terms, hierarchical structure and cross references; search formulations.

The factors that govern the structure of a file are therefore the use to be made of its data and the facility with which any one record or group of records can be addressed. Having considered how to access a field within a record, how do we cope with the problem of access to a particular record within a file?

In accessing any individual field, little account was taken of the medium on which it was held; nearly all manipulations of data elements within fields are carried out in the core store. (It used to be known as the Immediate Access Store, which described it adequately. It is known as 'core' because its principal electrical element used to be a wound core carrying positive or negative current.) The point about core storage is that any one site in the store is neither more nor less quickly accessible than any other. On the 'backing' stores – disc or drum or tape – this is not so. With a disc whirring round under the read-head, at any given time one segment of the disc is closer to the read-head and thus more immediately accessible than any other; on a tape, one record may be a hundred times more remote from a reading-head than another. In core, very small subdivisions of the total (words, or bytes, or slabs, or whatever other name the manufacturer has decided to call his smallest portions of core) are directly addressable and all subject to the same delay (virtually nil) in access. The data held on backing stores has to be brought into core before it can be processed.

So in considering how to organize records on the file, it is necessary to consider not only what classes of record can or should be grouped together because they are used together but what sort of access to them is required. Usually, these two considerations interact and have to be considered simultaneously; this makes planning a little more difficult. As far as possible in this chapter they are treated separately. (Commonly, groups of similar records form 'clusters' within a file and these headings of clusters can be first searched; then the records comprising the cluster can be addressed individually if the cluster heading indicates that a relevant record may be found within this cluster.)

Accessing records within a file

Three common ways of addressing records are serial, index-sequential and list, or list-processing.

Serial files hold records in the sequence of the 'key' – that unique identification number which we noted as essential on page 3. Records on tape are, by the nature of the medium, serially held, though normally 'sentinels' (quickly identifiable markers) are used at intervals along the file and the basic computer software allows the search to proceed from sentinel to sentinel without stopping to examine individual records, until the required sentinel is reached that precedes the group of records containing the sought record. Thus, for example, a set of document records in which a class number is given can be put on to tape with a sentinel marker at each change of class: for instance, if each has a five-digit UDC number, a sentinel can be put before each set of records beginning with the same first three digits. In the 'housekeeping routine' which does the scanning for a record, it is possible to start reading record by record only when the required sentinel has been reached.

If serial records are put on to drum or disc, an index can be made to say, for each segment of each track, which group of records it contains. When the file is updated by adding or deleting records, however, or by changing the length of a record, the whole

Storage: Systems, Files, Records and Fields 41

of the rest of the file has to be, as it were, moved up or moved down. This is easy on tape where normally records are transferred from the 'read' tape to a new 'write' tape every time a file is updated; it is more difficult on disc or drum where by adding a record in the middle of a file you may cause an overflow in that and all following segments.

Index-sequential. In this case, a common usage is the so-called index-sequential. In any one segment, the records are held serially, but the first record of segment number two is not necessarily that immediately following the last record in the segment number one. This was the case when the file was first set up, but overflows gradually occurred. When the file was set up, record BA followed immediately after record AQ; and since AQ came at the end of segment 1, BA became the first in segment 2. Now, however, there has been a massive updating of the file and many records intervene between these two; the solution is to put them into another segment, in this case segment 15, and make an index:

AA - AQ   Segment 1
AR - AX   Segment 15
BA - ..   Segment 2

Or one can put an indicator after one record to show that there is another record, serially between it and the next record in this segment, which can be found in segment nnnxxx. Several index entries do not need to be updated. All that is necessary is an additional 'line' in the index giving the address of the next space to be filled when an overflow occurs. On seeking record AF, the index directs the search to segment 1. A serial search through the records in that segment brings one to record AE which was followed when the file was set up by record AG; but a pointer at the end of record AE indicates that the next record (serially) to AE can be found in segment 33. Segment 33 can be a complete hodgepodge of records, but each can be accessed by only three accesses: (a) to the index; (b) to the segment indicated by the index; (c) to the segment indicated by the pointer within the indexed segment.

This idea of using a pointer to get to the next record brings us to list processing.

List processing provides a means of going serially through a file irrespective of where any individual record is stored. An index gives the means of going direct to an individual record.

Assume four records, A, B, C and D, of variable length, arriving into the file in the sequence C, B, A, D. C, being the first, is entered into location zero; it is 28 characters long, so it extends from 0 to 27; at the moment it is the only record in the file, so 28 has an asterisk entered into it, to show that this is the end.

Now B arrives. The first location available for it is 29, and it extends from 29 through to 51; in addition it has the extra character for the 'pointer', and into this (character 52) is entered zero to show that the next record begins at character zero.

Record A is the next to be input and obviously will be stored starting at location 53. It is nine characters long which (adding one for the 'pointer') means that it occupies locations 53–62. Into 62 must be entered the pointer to the next record in the sequence, the record for B, so it will hold the value 29 (29 being the start address of record B). Finally record D enters the file and is entered into the next available space, so it begins at location 63. At the end of its record will, later, be entered the correct pointer to record E; but at present D is the last record in the sequence and its pointer holds an asterisk to mark this fact. The pointer at the end of C must be altered from an asterisk (until now it was the last record in the file) to 63, leading from C to D. While the records are being added, an index to the file is being created in store; at the completion of the entry of record D it reads:

A 53 (its contents are in 53–61, and 62 holds 30, the start address of B)
B 29 (its contents are in 29–51, and 52 holds 0, the start address of C)
C 0 (its contents are in 0–27, and 28 holds 63, the start address of D)
D 63 (its contents are in 63–75, and 76 holds an asterisk)

To go serially through the file the program need look up only the starting point; thereafter each record indicates where to go for the next record.

This technique can be extended to give a sub-set of the file. Take an inverted file in which each record is based on the name of a descriptor. One will want to be referred from, for example, the R records to the S records, but also to the RA, RE, etc. records; Figure 6 shows how it is possible to move to the 'brother' S as well as to the 'sons' RA, RE, etc.

Storage: Systems, Files, Records and Fields 43

There is, of course, no need to limit oneself in the number of pointers. A system with links to 'father' as well as 'son' and 'brother' records could be used, for example, with a file of records hierarchically characterized. A record classified, for instance, 517.3 can have a pointer to its 'father' 517, its 'brother' 517.4 and its 'sons' 517.31, 517.32, 517.33, etc.

Inverted filing. 'Inverted' files are those in which the identification of any one record is the descriptor (indexing term or class mark): each record comprises the identifications of all documents having in common that one descriptor. The record for descriptor A is the identities of documents to which A applies, and a similar record is kept for every indexing term. Thus if the records for descriptors A and B each show documents 143 and 279, both these documents have been assigned (among others) to both A and B. A disadvantage of inverted filing is that in addition to the files carrying the indexing information it is necessary also to have another set of records giving the bibliographic data; to be told that numbers 143 and 279 are both relevant to an enquiry is normally insufficient – a means must be provided of finding out what these two numbers represent.

Hash coding. The expression 'hash coding' is pure jargon; in effect a hash code is a code meaningless outside the system in which it is employed. One way of using hash coding is to let the key number of a record be composed in such a way that from it can be generated the address at which it is stored.

For instance, suppose records to be of fixed length, forty characters, the first record beginning at character zero, the second at character 40, and so on. It is possible to arrange the files such that a rule for generating the address of any record is $40(n-1)$ where $n =$ the key number of the record. Updating is simple, the record goes into its correct slot, and subsequent accessing is simple also. The difficulties come where records are of variable length. This form of organizing access to a file, while it makes optimum use of the storage space and, indeed, is the only conceivable way of tackling inferential retrieval, does make demands on the computer, both for space and for time. Indexes, sometimes very long ones, have to be maintained, in locations which can be rapidly accessed, and extra programs are required for the maintenance of these lists

Address	Name	Link and address of next letter	Link and link address this letter	Meaning of last field
0	R	S 1		address of S
			RA 3	address of link RA
			RE 72	address of link RE
3	RA		RAB 196	address of link RAB
			RAC 207	address of link RAC
			RAD 128	address of link RAD
128	RAD		RADA 32	address of link RADA
			RADI 43	address of link RADI
43	RADI		RADIO 29	address of record RADIO
29	RADIO		RADIO* 1067	address of file RADIO*
			RADIOG 789	address of link RADIOG
			RADIOL 240	address of link RADIOL

6. 'Brother' and 'son' addresses in list processing

when records are added to or deleted from the files, or amended. This has to be weighed against the undoubted flexibility which the method permits.

A record cannot overflow into the area proper to another record, and when the records in the file are not in a sequence which permits generation of such an economic address as our example above, very inefficient use may be made of the backing store. If these limitations are borne in mind, hash coding can be a most effective way of disposing of and addressing a record.

Arranging files

Just because a collection of information about an entity is received as one record that is no reason for its remaining all together in one record during its lifetime in the system. Files are very frequently split up into other files for the convenience of processing; it is common for forty per cent of program-execution time to be concerned with sorting and merging in the process of converting one or more files into other files. As simple an operation as making a KWIC and author index, for instance, will:

make up a file of titles + document numbers;

make up a file of author names + document numbers;

generate KWIC entries, with the indexable word at the front of the record, on another file;

sort on the indexable word, putting the result into another file; go through that file, centring the indexable word, and printing the output file;

sort the author file into alphabetic sequence, and print it;

sort the document number file into number sequence, and print it.

What is to be done with a file has a great bearing on where it is stored. Material for sorting or merging is commonly held in serial form, on tape, since it has to be read serially during the sort/merge operation and tape is a convenient form of linear storage, with few 'housekeeping' problems which are obviously unwanted in a file that is to have only an ephemeral life. The main bibliographic records could be most conveniently held on exchangeable disc store: they take up a lot of room and should not, therefore, be loaded when they are not required, but should be available in non-linear form when they are in use. The information scientist designing an information system is advised to learn from his own computer staff the characteristics of the various types of storage available to him.

CHAPTER 4

Search Strategies: Retrieval

INTRODUCTION

All the search programs yet devised fall into one (or more) of three classes: serial-file searching, inverted-file searching and list-processed searching. This applies whether the system uses a highly structured thesaurus, a hierarchical classification scheme, or the free-est of free text. Each makes different demands upon the system design, and it is not possible to say whether one is better than another without detailed knowledge of the circumstances in which it is proposed to use them.

Even if an ISR system were no more than a simple search-and-print-out-the-result system, the files could not be organized solely according to the search criteria, and account must be taken of the balance of work between searching and updating; the other uses to which the available information may be put also make their demands on the total system design. This chapter is concerned *solely* with searching techniques. It should not be taken as a guide to the sorts of file structures required by ISR systems which additionally produce, for instance, KWIC indexes, and their brethren (KWOC, KWAC), classified listings, abstract journals, feature cards, dual dictionaries, and all the other tools. All these activities have their own requirements and influence the total system.

ENQUIRY FORMULATION

The enquiry logic used in an ISR system depends very much upon the retrieval language, but generally the relationships between the descriptors in an enquiry are the common Boolean relations 'and', 'or' and 'not'. In what follows the ampersand (&) is used for 'and' in logical formulae. There are other symbols (e.g. + for 'or' and . for 'and') but they are not standardized or always easy to remember.

Search Strategies: Retrieval

An enquiry for A & B or C & D could in fact mean any one of several different things:

it could mean (A & B) or (C & D) – example I
or it could mean A & (B or C) & D – example II
or it could mean A & (B or (C & D)) – example III

We can define these differences by saying that the various descriptors are on various planes or levels. In example I, (A & B) is on the same plane as (B & C). In example II, A, (B or C) and D are all on the same level while B and C are each on the next, lower plane. In the last example we have three planes: C and D each being on the lowest plane. It is essential to know beforehand what kinds of enquiry (in terms of number of descriptors and number of planes) are to be expected by the system. The planes must be clearly shown to the computer on input. There are various ways of doing this: parentheses can be used, as above, or the logic code can be put outside the parentheses so that its influence is supposed to cover all until the next parenthesis

& (A B or (C D E)) thus means A & B & (C or D or E).

With a more complex enquiry the lowest plane descriptors can be grouped into 'sets'. For instance, in an enquiry

(((A or B or C) & G) or ((D or E or F) & (H or I)))

sets can be made up thus

Set 1   A or B or C
Set 2   D or E or F
Set 3   G
Set 4   H or I

These sets can be grouped in concepts

Concept 1   sets 1 and 3
Concept 2   sets 2 and 4
Enquiry   concept 1 or concept 2

Another way is to number the levels

A_1 & B_2 or C_3 and D_{321} thus means A & (B or (C & D))

Personally I prefer the last; parentheses are easy for the enquirer but harder for the computer program to handle; sets and concepts are fairly easy for the enquirer but need a sub-routine to convert the groups of ideas into an enquiry in the form which the search program can take. But numbers are easy for the enquirer, need no conversion routine on input, and make for very easy search programs. However, this is a subjective decision; it is a matter of balancing human against computer effort and deciding which is the optimum in your own particular circumstances.

In addition to 'and', 'or' and 'not', there are two other factors to be taken into account. You may often want to qualify a descriptor by saying it must be 'between': for example, processes 'carried out between 1000° and 1500°'; or papers published 'between 1960 and

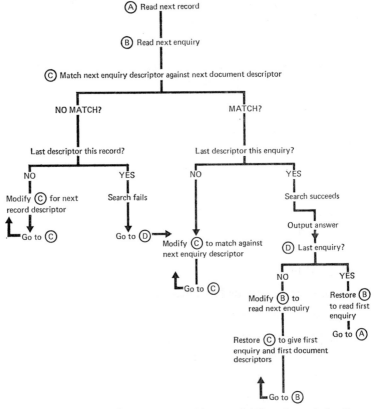

7. Sequence of operations when searching a serial file using only 'and' logic

Search Strategies: Retrieval

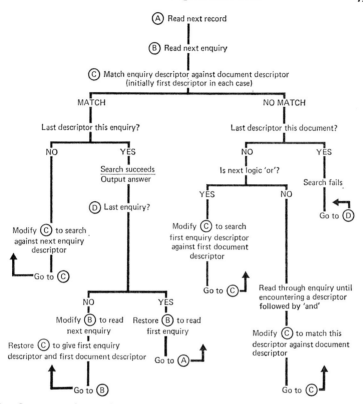

8. Sequence of operations when searching a serial file using 'and' and 'or' logic

1967'. The other qualification is 'out of': 'all students with any four out of the following eight "A" levels'. ('A' levels are an English term for school examinations of the highest standard.)

SEARCHING

Searching serially ordered files

In a serial file, each record represents a document, and contains at least its identity and the descriptors attached to it. Other information such as the title, or even an abstract, may be included according to the requirements which the system has to meet.

In a serial search, the first descriptor is matched against the first descriptor of the document; if there is no match, the instruction is modified to match it against the second descriptor and so on. For 'and' logic this is all that is necessary: when the enquiry descriptor has been compared with all the document descriptors and no match has been found, the search has failed. If there is another enquiry in this batch, the same process is repeated with that enquiry; if not, the next document record is read and the process repeated, starting with the first descriptor in the first enquiry. Figure 7 flow-charts the process. This flow-chart is, however, not sufficient for 'or' logic; in that case if the answer to 'Last descriptor this document' is 'No' the search has not yet failed – the whole process must be repeated for each of the 'or' descriptors and the flow-chart then takes the form shown in Figure 8. Even now the flow-chart does not represent all kinds of simple logic – for instance, if the logic were 'not', a match would result in the search failing, not succeeding; nor does this flow-chart provide for several planes (p. 47), or the necessary 'housekeeping' ('end of file', and so on). Figure 9 gives a flow-chart for 'and', 'or' and 'not' on several planes, using the numbering system. But it does not include 'between' and 'out of'.

One technique for dealing with 'between' is to put a symbol after the descriptor in the document record if the descriptor is, in fact, followed by a value (for, after all, it does not follow that a descriptor, which could have a numerical value attached, necessarily does so in every case). The same symbol follows the descriptor in the enquiry together with the two values between which the document descriptor value must lie. The program first seeks the descriptor and, having found it, the symbol; if the symbol is present it can then proceed to compare the values.

Another way of handling it is to use a dedicated (specified) part of the record only for that field, and enter only the value into that field.

'Out of' enquiries must include the words 'out of' (or a corresponding symbol) and the number to be found, then the string of descriptors from which the selection is to be made. For example 'OUT OF, 4, ENGLISH, FRENCH, GERMAN, ITALIAN, RUSSIAN, SPANISH, SWEDISH'. On encountering the symbol the program sets up a counter holding in this case the number 4. The various descriptors are then sought in turn and one is subtracted from the count when one of them is found. As soon as the count reaches zero, the

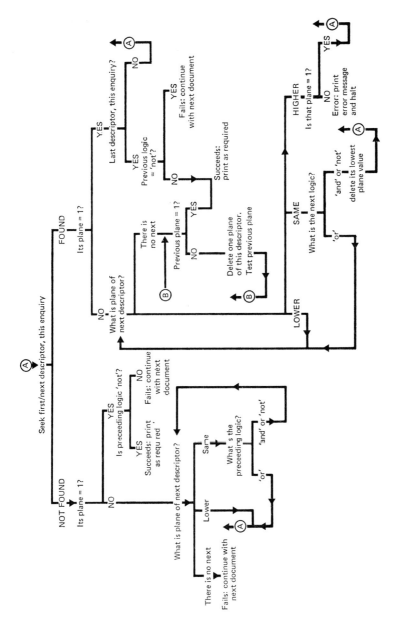

9. Sequence of operations for item search using 'and', 'or' and 'not' logic, with planes shown numerically

(a)	Working area	1	&	A
	Plane-1 area			
	Plane-2 area			
	Plane-3 area			

(b)	Working area	2	OR	B
	Plane-1 area		&	A
	Plane-2 area			
	Plane-3 area			

(c)	Working area	3	&	C
	Plane-1 area		&	A
	Plane-2 area		OR	B
	Plane-3 area			

(d)	Working area	4	—	D
	Plane-1 area		&	A
	Plane-2 area		OR	B
	Plane-3 area		&	C

(e)	Working area			EMPTY
	Plane-1 area		&	A
	Plane-2 area		OR	B
	Plane-3 area		—	(C & D)

10. Inverted file search for A & (B or (C & D))

Search Strategies: Retrieval

(f)
Working area	EMPTY
Plane-1 area	& A
Plane-2 area	− B OR (C & D)
Plane-3 area	EMPTY

(g)
Working area	EMPTY
Plane-1 area	A & (B OR (C & D))
Plane-2 area	EMPTY
Plane-3 area	EMPTY

condition is satisfied and the remaining descriptors need not be sought. If, after all the descriptors have been sought, the count is still not down to zero, the condition has not been satisfied.

It is obviously a good plan in serial searching to put the single, high-plane descriptors at the beginning of an enquiry. An enquiry reading (A or B or C or D or E) & F is going to be time-wasting if A, B, C, and D are sought in vain, E found at last and then F is not found. Another time-saving device is to use a table of frequency of use of descriptors, and search first for the least likely descriptor (if it is an 'and') or the most likely descriptor (if it is one of a string of 'or's). Such a table can be maintained during the input of new document information.

It is not, of course, necessary to store the descriptors in verbal form. They can be stored in fixed-field by giving a code number to each, or even by giving a dedicated bit position to each.

Searching inverted files

In an inverted file system, the record comprises the descriptor with a list of the documents (usually of the document numbers) to which it has been attached. If the system is to output anything but a list of document numbers, therefore, it is necessary also to have a file which can be accessed by document number, and which gives the additional information beyond the mere descriptors. (In this section, familiarity with the concepts of binary operations is important. They are dealt with in Appendix 2.)

The inverted-file record may contain the actual document numbers or the principle of the dedicated bit may be used, each bit position representing, in every record, one document only; a 1 or a zero in the bit position shows whether or not that document has been indexed with the descriptor in question. Even if the actual document number is used for storage purposes (to save space, but see below, page 57) it is not uncommon to convert it to dedicated bits while actually performing the search in core, to take advantage of the computer's ability to perform logic operations on bits at very high speeds. Figure 10 gives a simple example of the operation of this technique.

The enquiry A & (B or (C & D)) can be written as A_1 & B_2 or C_3 & D_{321}. In (a) the first descriptor has been read into the working area, with its plane and its logic operator. In (b) it has been read into the area for plane 1 and the second descriptor read into the working area. In (c) and (d) descriptor C, with its logic, has been read into the area for level 3, and D is already in the working area. (Note that at this point the absence of a logic symbol in the working area shows that the end of the enquiry has been reached.) The working area is on the same plane as a plane-area which has been filled: it is therefore necessary to combine these two. The plane-3 area shows 'and' logic so an 'and' is performed between the record in the working area and the plane-3 area, the result being left in the plane-3 area (e).

The end of the enquiry having been noted, all that is necessary is to work up the planes. The result of an 'or' operation between planes 2 and 3 is shown in (f), and of an & operation between planes 2 and 1 is shown in (g). The result is the answer to the enquiry in the plane-1 area, with the areas for the other planes empty.

The plane-1 area does not hold the actual answers, only the bits representing documents which do answer the enquiry. These bits can be counted and the total printed out; or they can be converted into document identity numbers and these printed out; or they can be converted into computer addresses from which fuller details can be obtained for printing purposes. In formulating the enquiry, as in this example, it is always as well to put the higher-

Search Strategies: Retrieval

The query is:
A & B & ((C or (D & (E or (F & G) or H))) & (I or (J & K)))

It can be expressed in terms of planes as follows:
A_1 & B_1 & C_2 or D_3 & E_4 or F_5 & G_5 or H_{432} & I_2 or J_3 & K_{321}

The process of operations in handling this query is as follows:

Read	Logical operation	Result	Test, and resultant operation
A→WA→1			
B→WA	& WA/1	1 holds A & B	If 1 is empty and its logic is &, FAIL
C→WA→2		2 holds C	
D→WA→3		3 holds D	
E→WA→4		4 holds E	
F→WA→5		5 holds F	
G→WA	& WA/5	5 holds F & G	
H→WA	or WA/5	5 holds F...H	
	or 5/4	4 holds E...H	
	or 4/3	3 holds D...H	
I→WA			WA has level higher than last-filled, so work up existing levels until level of WA (i.e. of I) is reached.
	& 3/2	2 holds C...H	If 2 has logic &, and WA has logic &, test if 2 is empty. If it is, FAILS. If it is not, there is still something to & with, so store 2 temporarily.
	WA 2	2 holds I	
J→WA→3		3 holds J	
K→WA	& WA/3	3 holds J & K	
	or 3/2	2 holds I...K	If 2 is empty, there is nothing to & with other level-2 expression, so FAILS.
store WA	& WA/2	2 holds C...K	
	& WA/1	1 holds A...K	1 holds the answer.

Key
WA = working area   → = moved into   1, 2 etc = plane 1, 2 etc, area   & WA/5
= do a logical 'and' between working area and plane-5 area.

11. A very complex enquiry handled by inverted-file system

level expressions in front of the lower-level and to put the shorter expressions before the longer. An outline of the operation of such a program is given in Figure 11.

'Between' is no real problem in an inverted system; where a descriptor has a numerical value, not one bit but as many as are needed to hold the maximum value, are allocated to each document; if, say, the maximum value were one hundred, seven bits per document would be required. In binary notation 1111111 means $64 + 32 + 16 + 8 + 4 + 2 + 1$; 100 in denary is expressed as 1100100 in binary. On reading the descriptor record into the working area, each 7-bit group must be compared with the limits given in the enquiry; only if it falls within those limits is a 1 entered into the document's bit position in the working area.

'Out of', however, presents real problems; all the possible combinations must be computed; then a logical 'and' must be performed on the descriptors for each combination, followed by a logical 'or' on the result of these logical 'and's, before the search can proceed. Indeed the probability of frequent 'Out of's in enquiries is a good reason for not choosing an inverted filing system.

Storage space in inverted filing systems. An objection to the use of dedicated space in an inverted file is the number of empty bits that will be found in every file, the heaviest-used descriptor seldom applying to more than a small fraction of the total descriptors. There are two ways of getting over this inherent waste of space. First, let the file record itself carry the full numbers, converting them to bit positions only when the record is brought into the core during search. Second, instead of a string of empty words representing a whole string of documents to which this descriptor – the subject of the record – does not apply, a word differently signalled can give the number of words which would have been used if every single document were allocated a bit position. So each word would carry as its most significant bit (normally that on the extreme left of the word)

either (a) a signal that the bits in the word represent documents;

or (b) a signal that the bits in this word represent a binary number, the number of words that would be utilized if every document were allocated a bit position.

Search Strategies: Retrieval

The rest of the bits comprising the word would

either (a) represent documents;

or (b) represent a binary number.

Suppose that the first three words of a record for one descriptor read as follows (assuming here an 11-bit word):

word 1   1 0 0 0 0 0 1 0 0 0 0
word 2   0 0 0 0 0 0 0 1 0 1 0
word 3   1 0 0 0 0 0 0 0 0 0 1

The first (most significant) bit in words 1 and 3, being a 1, indicates that each bit in the word represents a document; so from word 1 it is known that document no. 6 is indexed by this descriptor. Word 2 has its most significant bit set as 0, which indicates that it represents words which would be full of zeros: each word would hold ten document-representing bits, and since the binary value represented by the bits in this word is ten, the whole word carries an indication that documents numbered 11 to 110 are not indexed by this descriptor. Word 3 carries a 1 as its most significant bit, so its successive bits (again we read from left to right) represent documents numbered 111, 112, 113, etc. The bit representing document number 120 is set as 1, so document 120 is indexed by this descriptor. It has only taken these three words to represent 120 documents of which only two are indexed by this descriptor. Since any one descriptor applies only to a very small percentage of the documents, some form of record compression such as this is valuable.

Searching List-Processed Files

A simple application of list processing is illustrated in Figure 12. In this figure are shown only the first two descriptor fields of some records on a file; there will be other records between those shown, but here are given only those to which descriptor A or descriptor B apply. A search for items with A & B starts its detailed examination with record 17, which is the first to be encountered, in going serially through the file, which holds A. Since the search is for A & B, the program examines the next descriptor field of this record to see if it is a B: it is, so this record matches the search pattern.

Record number	First two descriptor fields	
	Contents	Meaning of contents
17	A	Descriptor name
	21	Next record with this descriptor
	B	Descriptor name
	146	Next record with this descriptor
21	A	Descriptor name
	99	Next record with this descriptor
	G	Descriptor name
	735	Next record with this descriptor
99	A	Descriptor name
	196	Next record with this descriptor
	L	Descriptor name
	1326	Next record with this descriptor
146	B	Descriptor name
	196	Next record with this descriptor
	G	Descriptor name
	735	Next record with this descriptor
196	A	Descriptor name
	735	Next record with this descriptor
	B	Descriptor name
	426	Next record with this descriptor

12. List processing: five records from the beginning of a list, giving for each only the first two descriptor fields. Records 17 and 196 hold both A and B, so from record 17 the program will go to 146 and thence to 196.

The record shows that another record with A is record 21, but 21 cannot hold A & B, since record 17 shows that the nearest record having B is 146. The program therefore can go direct to 146; this record is known to hold a B, so all that is needed is to look for the presence of A, too. A is not there, so the program can go direct to the number given as the next record holding B, in this case, 196. Here, too, all that is necessary is to look for A, and in this case it, too, is present, so 196 matches the search requirement. It gives the numbers of the record next holding A, and of that next holding B, being 735 and 426 respectively. There is no point in looking at 426, since it cannot hold both A and B, so the search proceeds direct to 735. And so on to the end of the list. Thus, out of 735 records, to

Search Strategies: Retrieval

look for those with A & B, only records 17, 146, 196 and 735 need detailed examination. A general purpose program such as this can save a great deal of search time, though it is not a simple program to write.

To maintain this file it is necessary to have an index giving the last-used address for each known descriptor. When more records are added, this index must be updated to give the last-used address in respect of every descriptor added to the file.

Some optimization of this system can be obtained by using what is sometimes known as a 'page' – a page being a segment of store. The file of records is divided into such pages, and a page index is kept for each as in Figure 13. In this example, descriptor A is

Keyword	Page	1st record on this page	No. of records on this page
A	1	7	4
	4	3	2
B	1	4	3
	3	2	2
	4	2	5
C	2	4	2

A & B Examine three B records on page 1 for A
Examine two A records on page 4 for B
i.e. only five records to be accessed/searched, on all these three pages

13. Keyword 'page' dictionary

shown as occurring in records held on pages 1 and 4. On page 1 it occurs first in the seventh record and on page 4 in the third. There are four records on page 1, and two on page 4, which contain descriptor A. Similar summarized information is given for every descriptor.

Assume an enquiry for descriptors A & B & (C or D). An initial scan of the first column of the index shows that only pages 1 and 4 contain records with both A and B, so only these two pages need be searched further. Page 1 has four records containing A and only three containing B, so naturally search is directed to the

B records, each of which will have a link to the next record on that page containing a B.

On the other hand, page 4 has only two records holding A and five holding B; in this case the search goes direct to record no. 3, which is the first of the two records on this page of which A is one descriptor, and there looks for B and (C or D). This done, it then jumps direct to the next A record. In fact, it may not be necessary to examine this second A record at all: if the link address of the B descriptor (if any) in record no. 3 is larger than the link address of the A descriptor then (since there are only two A records on this page) further search on this page is unnecessary.

By using this descriptor/page dictionary it has thus been possible to reduce the total number of records which must be examined in detail to at most five: three from page 1 and two from page 4.

Depending on the type of backing storage used, it may be desirable to cut down the number of movements of the read-head; this is particularly of value in a real-time system. A continuous sweep of the store for all the pages required can be achieved by creating an inverted file of all the pages to be searched (Figure 14).

QUERY	PAGES containing relevant records
Q1	1, 5, 12
Q2	4, 12, 19, 25
Q3	1, 19

After page sequencing:

PAGE	EXAMINE FOR
1	Q1, Q3
4	Q2
5	Q1
12	Q1, Q2
19	Q2, Q3
25	Q2

14. Multi-processing with list structure

Page 1 is accessed; the list for enquiry 1 is retrieved and processed and the list temporarily stored. Then the list for enquiry 3 is similarly dealt with. The read-head can meanwhile be moving on to page 4 for its delivery into store and examination for enquiry 2. As new enquiries come in, the program looks up the descriptor/

Search Strategies: Retrieval 61

page index and the enquiries are entered into the page list, so that a continuous scan is possible.

'NATURAL LANGUAGE' SEARCHING

A discussion of the pros and cons of using the words in the text as opposed to words assigned from a controlled vocabulary, in order to represent the text in the computer, is not primarily within the scope of this volume of the series. Whether an author's own words make better index concepts than those attributed by indexers, or deduced in various ways from the text, is immaterial here; what does matter is the way the records and files are organized for retrieval when there has been no formal indexing by a human indexer.

It should therefore be said at the start that 'natural language' systems do not differ, in their searching techniques, from those already outlined in this chapter. The sentences as input are transformed, somehow or other, from a string of characters into an organized record. The result of the transformation may be a concordance (the simplest form of inverted file), a serial file with descriptors of some kind extracted from or deduced from the text, or an inverted file using descriptors extracted from or deduced from the text. In the course of this transformation, dictionaries held within the computer may be consulted by the program to index with a preferred term a synonym or near-synonym of it occurring in the text.

The frequency of occurrence of the words, either alone or in combination with other words, is often measured to indicate their value as descriptors. In brief, the text is transformed into records which can be dealt with for searching in the same way as described in the earlier parts of this chapter.

An imaginary system

The following description of a notional system combines many of the features of natural-text searching systems and can serve as an indication of the type of system that is required.

The text is transcribed from source or abstract to machine-readable form and during reading the process of content transformation can begin. First, each word must be checked: it is read and its location in the text is noted and it is then looked up in a master dictionary which merely lists every word that has entered the system; words that are not found in it are either mis-spellings or new words.

These unmatched words are printed out with their contexts (twenty words on either side) and locations. Spelling and punching errors (which look like new words) are corrected and re-input. For the genuine new words a human decision is made:

(a) is it a term in its own right, or a synonym?

(b) if it is a synonym, of what?

(c) if it is a term, does it involve automatic posting on to a higher ('father') term?

All these are fed back to the machine for the 'edit' run. In this run, the mis-spellings are corrected; the master dictionary is updated with the genuine new terms and, at the same time, a 'concordance' of the document is made, listing every non-function word (i.e. every word other than 'the', 'of', etc.) and adding, if a word is not new, its address in the term dictionary. New words are entered in the term dictionary which carries the term, a count of the total times it has been used in the entire data base, and the address of its inverted file(s). (If it is a synonym, the inverted file address will be that of the term which is used for it.)

Weighting of terms. When the document is posted to the inverted files, a weighting is calculated – the weight for the term in this document. This decimal fraction is:

> the frequency of the term in the total collection *multiplied by* the frequency of the term in the document (normalized to a common denominator since documents are of varying lengths) *and the result divided by* the total number of terms in the collection.

This weighting is stored, alongside the document number, in the inverted file (see page 65).

There is one provision here which partly deals with one of the bugbears of 'natural-language' systems: where a word is a homograph (and noted as such in the master dictionary), the weighting is reduced by an amount determined by the number of homographs (the number of different meanings of a single word).

Search Strategies: Retrieval   63

The entries in the term dictionary carry cross-references, so that on demand a term list can be printed out giving the absolute frequencies of each term, with the words it has been used for, and the broader ('father') and narrower ('son') terms of each. This too can be called as a display: a request for a word in the master dictionary will display the term to which it has been posted and the other words and terms associated with that term.

Searching is like that for any search on weighted terms. Boolean logic is used, so that a document is rejected if none of a string of 'or's is present, but all terms are sought and their weights summed for the document, giving a ranked output. This is for ISR; with SDI the number of documents is relatively small and the search is non-Boolean: the documents are sequenced merely by the number of profile words found in the document, with a weighting threshold to serve as a cut-off point.

Citation searching

It was noted on page 5 that citations of relevant works are among the information about a document which may be stored, but so far no reference has been made to utilizing them. Citation searching makes use of the probability that if A cites work X, and B subsequently cites X, then there is some relationship between the work reported in A and that reported in B. If several citations are common to A and B, there is an even greater probability of A and B having much else in common. Commercially available citation indexes (published by the Institute for Scientific Information, 325 Chestnut Street, Philadelphia, Pa., 19106, USA) make use of this principle, and it can be used for computer searching.

File organization is normally based on the author's name; subfiles are made for each of his works, and for each of these are listed the works which cite it. Every file must be capable of being added to every time the work is cited, and care is needed (unless the citing works are referred to by codes, with a look-up table to translate these codes) in expressing all citations uniformly. It is possible, too, to organize the file for retrieval primarily on descriptors, using citations as additional descriptors, or merely adding, on print-out, a reference to the fact that the retrieved document does or does not cite the work or works specified in the enquiry.

PRECISION AND RECALL DEVICES

Techniques for improving the precision or recall in searching are primarily connected with the indexing attached to the documents or with the form in which the enquiry is posed. Both are dealt with in other books on ISR but mention is necessary since they affect the computer system required for searching.

Links and role indicators

Links are used as symbols attached to the descriptors of a document to indicate that they are related. An item indexed by the descriptors 'Cleaning' 'Catalysts' 'Production' 'Sulphuric acid' (being about the cleaning of the catalysts used for the production of sulphuric acid) might be retrieved for any enquiry about cleaning and sulphuric acid; a symbol (commonly numeric) attached to 'Cleaning' and 'Catalysts', and a different symbol attached to 'Production' and 'Sulphuric acid' could, as it were, tie these four descriptors into two groups, preventing such an error.

With item filing, such symbols must be attached to the descriptor within the item record, and for ease should be placed in the first character. Then, in the supposed query above, while the program would look for 'Cleaning' without being concerned about its symbol, it need only examine a descriptor for match to 'Sulphuric acid' if the correct symbol were also present. With inverted files, the link has to be carried around with the document number.

Role indicators show not merely that there is a relationship, but also what that relationship is, distinguishing, for instance, between 'Destruction by bacteria' and 'Destruction of bacteria'. The symbolism is more complex, being two-part at least: one part to show that these two words are linked and the other the nature of the linkage. It is possible to add these selection criteria to the search algorithm without forcing it out of the basic serial- or inverted-file search.

However, attempts are being made to devise a symbolism that will give a complete context to any individual descriptor, so that it might, perhaps, be shown as related to a number of other descriptors (for the one document) each in a different way. With existing computing techniques there is no other way of handling this than the building up of a connection table (similar to that used in repre-

Search Strategies: Retrieval 65

senting the structure of molecules). To effect retrieval with this kind of structure involves an increase in complexity of both the record structure and the search procedure, and there are sacrifices in speed of searching, all to be set against the benefit of precise retrieval. So far as I know, it has not been attempted in a working system.

Weighting

An example of one use of weighting is given on page 62. Where weights are attached to the descriptors of the document the search algorithm can search for all the descriptors of the enquiry and compute either their total weight or their average weight. The enquiry may specify a threshold value, which this computed value must attain if the document is to be retrieved. Provision must be made for storing the weighting for each term for each document against the term in an item file, or against the document number in an inverted file; but the program is only a little more complex than that required for the unweighted search. A weight can be also attached to the enquiry descriptor, to be matched against the weight attached to the document descriptor, before the decision is made whether a match has been effected between these two descriptors. The retrieval program becomes somewhat more complex, but is the same in kind as the unweighted search.

A different search program is required where weightings are given to the descriptors of the enquiry but not to those in the document records. An example will clarify the method. An inquiry is given as being for:

A – weight 10
B – weight 10
C – weight 4
D – weight 3
E – weight 2

The threshold specified for this enquiry is 22 (out of a maximum 29). The effect is an enquiry for A & B & (C or D or E), but it allows the program to indicate in the output those items with weight 29 down to 22, i.e. it permits a ranking in a sequence predetermined by the enquirer (as opposed to those programs which, in the same way, search for every descriptor but merely output in the sequence of the number of descriptors matched). Record and file structure are as for Boolean searching, but the search program, having only 'and's plus a very small computation, is simpler and slightly quicker in execution.

Generic/specific ('father'/'son') relationships

Where there is a structured indexing language, ability to search on more/less specific terms than the one posed in the initial enquiry is used to improve precision and recall. No computer assistance need, of course, be necessary: when inputting a document's descriptors, it is simple enough for the human indexer to add where necessary the generic terms to the specific terms already indexed. This can, however, be a computer function, with a look-up table of terms giving the additional terms.

An inverted file of UDC (or other hierarchical classification) numbers can be used, each document number being initially posted on the most specific level possible for each descriptor. The program must be such that a search for 669 will retrieve documents classed 669.1, 669.2, etc.

Word truncation

It is, of course, possible to look for fragments of words stored in descriptor fields in an item file, but this is not an economic means of searching. It is more common to use an inverted file with a technique for addressing the required truncation similar to that described on page 58. A document number is posted once only, on the full term; access to all documents indexed by a word beginning with RADIO is made by first processing the list as far as RADIO and thence obtaining the addresses of all files to which this is the prefix.

Context

In natural-text-searching systems where there is no vocabulary control, precision is often obtained by using contextual constraints. For instance, the words 'information' and 'retrieval' might be specified as being up to two words apart; this would allow the phrase 'information storage and retrieval' to be accepted. It may also be desired to specify that words occur in the same sentence,

Search Strategies: Retrieval 67

paragraph or chapter. This is achieved by storing, in the record for each word, (1) the identity of the sentence in which it occurs and the address of the record relating to that sentence; (2) the addresses of (a) the word immediately preceding it and (b) the word immediately following it. These three elements must be present for every occurrence of the word in the text. The sentence record similarly identifies its location within a paragraph, the paragraph its position in the chapter, and so on. It is therefore possible using list processing to make a kind of serial search without actually examining, at least once, every record in the file (or designated sub-set of the file).

CHOICE OF STRATEGY

It is impossible to lay down a hard-and-fast rule and say that one particular search strategy is better than another: too much depends on the circumstances. The search system requirements are influential, but even they are not over-riding. The search is only a part of the system, and it is system efficiency, not merely search efficiency, which is sought. The other major components of the system are the input of new information and the maintenance of the files of information, and the question every system designer in the field of ISR has to ask is, 'Am I making the search process fast and cheap at the expense of slow and costly file updating and maintenance?' The answer, of course, will depend on the relationship between the number of new items coming in (and therefore the extent of the file-maintenance programs) and the volume of enquiries. In addition, the numbers of descriptors per document, and documents per descriptor, influence the strategy to be chosen, and so does the kind of enquiry logic to be expected. It should be understood from the start that it is only the information scientist who can supply the data on which a sound decision is to be based. In many cases, he is not able to provide measurements which can be used; but he has (or should have) a sure 'feel' for the nature of the problem. And though qualitative rather than quantitative, his 'feel' will have to be expressed in quantitative terms to be used during system design.

However, some simple comparisons between serial, inverted, and list-processed file searching can be made and are of use (Figure 15).

Serial filing	Inverted filing	List-processed filing
Increase no problem: next record is merely placed after previous record	Increase means adding to as many records as there are descriptors to the document	Increase requires access to and updating of index as well as merely adding new record
Reference data can be held in the same store as the descriptors	Must cross-refer to another part of the store holding the reference data	Reference data held in same store as descriptors
Record normally unchanged after entry	Record subject to continual updating	Record accessed when its descriptor is next used in another document; no other updating of record
Change of descriptors for one document easy; only that record need be located	Change of descriptors for one document involves changing as many descriptor records as there are descriptors for that document	As for serial filing
Major vocabulary change harder	Major vocabulary change easier	Major vocabulary change easier than serial but harder than inverted
Every record must be examined for every enquiry	Only records of descriptors concerned need be used	Only probably relevant records need be searched
Enquiries can be batched	Enquiries can only be dealt with singly	Suited to batch or on-line
Man-machine communication not easy	Man-machine communication easy	Man-machine communication easier than for serial filing

15. Relative merits of three kinds of file design

Search Strategies: Retrieval

PROMPTING SYSTEMS

Prompting systems are used in all on-line systems to facilitate the accurate use of descriptors, both by indexers and by enquirers, to assist in the logical formulation of the enquiry, and to help the user make better use of his time by using the programs correctly.

Selection of descriptors

Prompting systems which aid the selection of descriptors have a bearing on the choice of search strategy, since they affect the kinds of file that must be maintained, and the accesses to them.

In a structured-language system, choice of descriptor can be helped by displaying the part of the thesaurus surrounding the term that the user thinks of using – the broader ('father'), narrower ('son') and related terms, as well as the synonyms and co-ordinate terms. (This can, of course, be done by supplying the user with a printed copy of the thesaurus, or getting him to consult a member of the information staff; the cost of lengthening the program to give this facility must be balanced against the undoubted value to the user of being able to formulate his own enquiry.) Where the vocabulary is unstructured or non-existent it is helpful to output the six most recent titles in which the proposed descriptor appeared, giving the actual descriptors attached to each document. This is a helpful way of bringing to the mind of the user possible associated keywords that he may wish to use.

In either case it is necessary to hold the thesaurus in alphabetic sequence. With a structured language, the record for each descriptor must give the related terms and their addresses. Where the language is a lattice rather than hierarchical, it is necessary to show the relationships between the broader and narrower terms. The sub-records for a single term N are indicated in Figure 16. N is an intermediate term in two trees (as often happens in lattices):

A narrower term   N   narrower term B
X narrower term   N   narrower term Y

The record for N must show its relationships to A, X, B and Y. At its beginning the record will hold data about N itself, its frequency count, for instance, and the address of its inverted file if any.

There will follow four sub-records, one for each of the terms A, B, X, Y, with a link to show the relationships:

Serial number of sub-record	1	2	3	4
Name of descriptor	A	X	B	Y
Address of this descriptor's record				(address given in this field)
Link address	3	4	1	2
Relation to N	H	H	L	L

NOTE. Here H represents 'higher than term of main record'
L represents 'lower than term of main record'

16. Four sub-records of one record of a descriptor showing lattice relationships

The 'link address' given in the fourth field of each sub-record shows the relationship between A and B and between X and Y. So if generic posting is being used, Y can be traced up through N to B. In the case of the higher terms, it is also possible to have a further field in their sub-records showing that still higher posting is required, in which case the information can be got from the records for A and X. The record as shown is sufficient for displaying all the relationships between N and its closely related terms; the actual addresses of the records for A, B, X and Y permit rapid access to their records for further display of relationships.

The record shown in Figure 16 can also have added to it the address of a short inverted file carrying the six most recently accessioned documents using this term, whether in their title or in the descriptors attributed to them.

Another form of assistance is an indication of the number of items retrieved by the use of the terms chosen. In its simplest form, all that is done is to give the user the actual number of documents on which the descriptor has been used. Or a simple algorithm can be used, based on the probability of occurrence of each descriptor, thus: suppose the search is for A & (B or C) & D and it is known that A occurs in 10 per cent of documents, B in 2 per cent, C in 5 per cent and D in 1 per cent. Then it can be calculated that in a collection of 30,000 documents:

A occurs in 3,000

B occurs in 60 and C in 150 of these 3,000

D occurs in 2·1 of these 210.

(It must be assumed in this that occurrence of any descriptor is

Search Strategies: Retrieval

completely random so that occurrence of A and B together is neither more nor less likely than their occurrence separately.) Formulae of greater or less complexity have been reported in the literature and can be found summarized in volumes of the *Annual Review of Information Science and Technology*.*

A different form of prompt routine uses the techniques of programmed learning to teach the user how to set about formulating his enquiry. Some interaction between the computer and the user is inevitable in on-line searching: the user has to identify himself to the computer and state which file he wishes to access; the machine has to check that he may in fact access this file and may also have to set up a routine for counting the time which he is taking and charging it to the right account. In addition it will confirm to him that the file is now available for his use, and may normally ask via the typewriter, 'Have you used this system before? Type Y for Yes and N for No.' On receipt of the response N, the user can, via the typewriter, be taken step by step through the formulation of his enquiry. For instance, the next question might be, 'Which part of the file do you wish to search?' Possible sections are . . .' and this can, when answered, be followed by, 'Please type in the name of the first descriptor you have chosen,' and so on.

THE REST OF THE SYSTEM

A surprisingly small amount of the total computing time in an ISR system is spent on searching, even after one has allowed for the time spent in extracting files from the backing stores and bringing them into core.

Most of the computer's time is spent on rearranging data. The form in which the data is originally input is influenced by human and material factors outside the computer itself. Even if it is typed direct to magnetic tape, it may well be that it is not in the form in which the computer can immediately operate upon it; some items for a single record may, for instance, only be available at a different time from the other items. The form in which data are stored for

---

* Cuadra, C. A., ed. *Annual Review of Information Science and Technology*. American Society for Information Science, vol. 4, 1969, pp. 34–36; vol. 5, 1970, p. 307. These make a good starting point for a search for further work on this subject.

one activity (searching, for instance) is often unsuitable for another (preparation, for instance, of a classified bibliography) in which case some sorting and editing are required. There are many occasions upon which editing, sorting and rearranging functions have to be taken into account. The form of output again often requires a compromise between what is best for the user and what is most convenient for the program. While it is true to say that reformatting a record according to the most bizarre requirements is perfectly possible, there is a cost to be paid in the amount of programming required to reformat the data and often in the running costs, since the data might be unsuited to storage in the form in which it can conveniently be printed in the desired format.

A program is not, in fact *a* program: it is a collection of programs most of which are concerned with reformatting or rearranging fields within records, records within files, and files within the store. Sorting and merging are the bases of all file transformations (Figure 17). If an inverted file is being created for newly received documents, for example, the following steps at least are required:

(a) read the input record for each document, deleting index fields used only in input and adding index fields used in the main document file, for processing; repeat for all input documents, adding the revised document records to the main document file, and updating the index to that file;

(b) for the first document record, make a list showing each descriptor and its document number; repeat till all newly input documents have been handled, making a file of descriptors for all new documents;

(c) sort the newly entered keywords on this file, into the same sequence as the main keyword file sequence (normally this is 'put the new keywords into alphabetic order'); where a keyword has been used in two documents, merge its two records into one. Make a file of these 'new documents' keywords for use in the next operation;

(d) take the first keyword record on this file; find the record for that keyword on the main keyword file and add the new document numbers to it; if no such keyword has been used before, make a new record for it on the main keyword file; repeat this for every word on the new input file.

Search Strategies: Retrieval

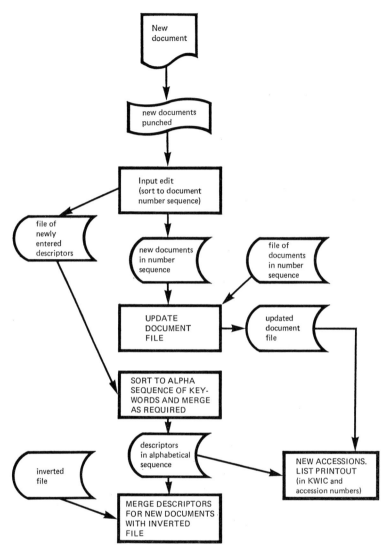

17. Flow-chart showing that most programs are sort and/or sort/merge

Note that all the work described here is sorting and file updating. Such operations depend upon the clear identification of the keys upon which these operations are to take place.

Most computer installations have at their disposal a number of slightly different sort programs and merge programs, which vary according to the type and location of the key. These handle sorting records within a file; the sorting of fields within a record is specific to the task in hand and package programs are normally of no avail.

In both cases, however, the program depends upon clear identification of the 'key' (or 'keys') upon which the operation is to take place. It must be known how many characters are to be compared in determining the sequence, and where those characters are to be found, and in what sequence significance is to be given to a set of characters.

Minutiae of this kind are often overlooked during system analysis, largely because the user has not given prior thought to the need for definition. It is important to realize that, however skilled he is, an analyst is at the mercy of the user. He is skilled at looking at the implications of what he has been told and at asking questions to elucidate them, but the user, unless he has himself taken the trouble to identify the details of the system, may well fail to put the analyst on the right tracks.

The analyst's aptitude for getting 'into the heart of' other people's problems is essential for every part of the project, but the person who will have the ultimate task of designing the system should participate fully in these earlier stages. He will be particularly useful at the stage of noting what data are available for the use of the proposed system, and what will be the outputs of the proposed system.

The result of the initial system study is an outline of a system showing how the available information can be handled to produce the required information. It is not necessary at this stage to go into details of how a process is to be performed, but it is necessary to show the principal transformations that will be required. It is at this stage, therefore, that critical decisions such as the nature of the indexing language should be settled, and the arguments marshalled for and against each possibility should be presented to support the decision made.

CHAPTER 5

Conversion to the Final Design, and Implementation

INTEGRATING THE PARTS

By now we have achieved a good picture of the plan, its implications, when it is to be produced, from what, and how. In fact, it is a series of designs of parts of the total system. Now they have to be integrated into a whole. It is at this stage that the final and definitive decisions are to be made on what functions are to be incorporated from the start, what is to be introduced at later stages, and when. It is now possible to make 'final' drafts of the input and output records and of the essential internally-stored files. A flow-chart with the suggested inputs, records and outputs is made for each sub-system but first a major flow-chart must be created for the whole system; gradually this is broken down into smaller (in coverage, not necessarily in size or in the quantity of information handled) and more detailed charts. This collection of flow-charts must show the movement of every single item of information from one place to another as well as the movement of large chunks of information (whole records or substantial parts of them). The preliminary draft must be accompanied by specimen input and output forms and the suggested types of storage formats, and the quantities of each. It is important to show at this stage what internal records are required to be formed. (There may yet be others, dictated by the computing requirements, but the time for these is not yet.)

The charts and forms must cover the extra- as well as the intra-computer files and formats. The people who will be handling these are an essential part of the system, and all parts of the whole system seen by human eye or touched by human hand must be represented.

The hardest part of all this is the complete integration of all components in a single whole. Hitherto this has been easy, with a stated or understood appreciation that A is a result of operation V, and that B is produced by operation W and is used for C in operation X and for D in operation Y as well as for final output as part of E. All this must now be made explicit and the processing required firmly understood.

These processing operations are of course indicated in the flowcharts but need expanding to clarify with what, when and how they are to take place. Both human and machine limitations must be allowed for. Very often the latter at least will have been predefined in general terms – that such-and-such items of equipment are going to be available, their capacities being such-and-such at a cost per day (or hour or minute) of such-and-such. But the manpower required is seldom specified and the requirements in terms of skills and times must be set out in detail in this first description of the system. It is not sufficient to say that the input is to be on punched cards laid out in a particular way: what is needed is to say what kind of worker prepared the material from which the input is to be copied (punched); how long it takes them and the (punch) operators to prepare one item; how often it must be done and hence how many such people/cards will be required for every input operation (on average). From all these data can be indicated the cost of preparing one such item of input. The output is to be considered in as much detail: how it is to be copied, transmitted to whom, how and where it is to be filed, etc.

Machine considerations

For every operation, it is implicit in all this that consideration must have been given to the equipment to be used, its capital and running costs (expressed in terms of the work proposed to be performed upon it), its effect on such things as input and file design, frequency of use, and so on. For every machine operation (whether computer- or humanly operated) it must be explained why it is proposed that the machine be thus used. Here the machine limitations must be made explicit; where alternative methods are possible, each must be listed together with its advantages and disadvantages.

It is in this preliminary design that the 'Why?' of every 'How?' must be made explicit. It must however be remembered that this is still not the final design, though it is the stage at which system

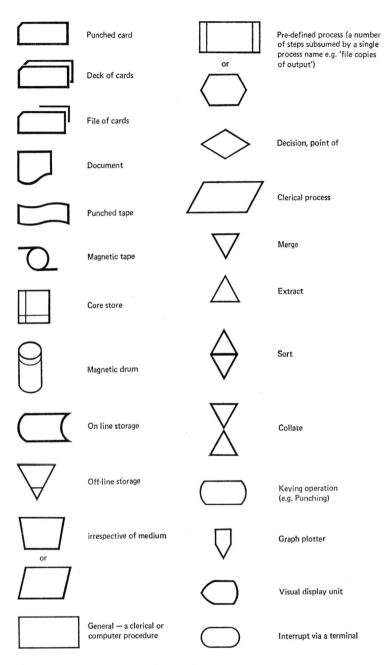

18. Some flow-chart symbols used in input or output, and storage

78 *Systems Analysis for Information Retrieval*

designer and analyst have their final chance of influencing others' decisions on what is to be made available. At this stage they are doing no more than stating what might (and in their opinion should) be done at what cost and utilizing such-and-such facilities in terms of staff, time and equipment. The decision on which facilities are to be made available is not theirs to make, though it is most emphatically their province to provide the information on which such decisions are to be made.

FLOW-CHARTS

The symbols used in flow-charting have not yet been wholly standardized, but a British Standard, *BS 4058*, does in fact exist: *Data processing flow chart symbols, rules and conventions*, 1973. There are still slight differences between the symbols used by some of the computer manufacturers; and those in figure 18 are, some of them, unique to one computer series; fortunately most people recognize (and often use) all the symbols.

Connection lines ('Connectors') in flow-charts. It is convenient for charts to flow from top to bottom of the page; but this is not always possible. For instance, from a 'decision point', after a set of circumstances has been tested, the program sometimes has to repeat some earlier steps. Where these are shown on a single page of the flow-chart it is often possible merely to draw an arrowed line connecting the two parts of the chart; and the same device can be used to show a jump forward. In these cases labelled rings are used as in Figure 19. Backward movement, to an earlier point, is

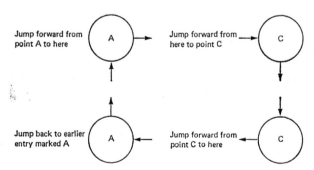

19. Symbols used to show jumps in flow-charts

Conversion to the Final Design, and Implementation

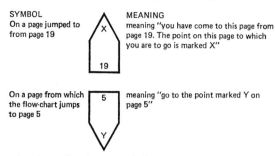

20. Arrow-heads used as jump symbols

commonly represented on the left, and forward movement on the right.

The 'arrow-head' can be used for much the same purpose. It is usually placed prominently at the foot of the page to which the jump is made.

The sign § (in the U.K.) or # (in the U.S.A.) is commonly used for 'number'. Thus §5 means number 5.

Since symbolism of flow-charts is not wholly standardized, it is as well to ask system designers and programmers for a list of the symbols they will be using, with the meaning of each.

Versions of flow-chart and system description

Eventually there will probably be two versions of both system description and flow-charts. One, for the users of the system, is much less detailed, the flow-charts showing only the main flow of operations and including extra-computer operations; it will incorporate such instructions as how to handle the flow of work and how to complete input documents. The other will be used for program maintenance and every detail of intra-computer activity has to be made explicit, including all those items carried within a record or a file merely for computer use. (These need not be shown in the *Users' guide*). Usually called the *Operators' manual* it also gives the operating instructions which tell the operator how to proceed in all likely (it is hoped in all possible) circumstances.

INPUT, FILE AND OUTPUT DESCRIPTIONS

These will be nearly identical with those described in Chapters 1 and 2 as far as the *Users' guide* is concerned. The various pieces of coding which have been added by the programmer to ensure perfect definition of every field, record and file are needed only in the *Operators' manual*. In both, however, it is necessary to show not only the size and nature of every field, but what it is used for, when and how.

PRELIMINARY SYSTEM AGREEMENT

The outline system must now be submitted to its future owners/operators. There is a great temptation for the system designer and analyst at this stage to inhibit any change to what they are putting forward. This they must firmly resist, leaving the client a free hand, though never failing to discuss fully the implications of all suggestions made, in the light of:

>effects on the overall efficiency of the system, especially the effects of changes in one part on changes in another part;
>
>effects on the morale, efficiency and costs of staff operating the system;
>
>effects on computer efficiency;
>
>effects on overall elements of costs or on parts of them.

All these must be taken into account in drawing up the final preliminary agreement.

The ultimate Preliminary System Agreement should comprise:

>a description in detail of the present system (if any);
>
>a statement of the objectives of the proposed system, with a time-table of their proposed implementation;
>
>a verbal description of the new system, including its breakdown into parts and their inter-relationships, and the points at which additional sub-systems will be grafted on to it;
>
>flow-charts of the new system and its sub-systems;
>
>detailed specifications of every input/file/output;
>
>costs, in terms of capital and running expenditure, of staff requirements, machine requirements, implementation (including programming, program testing) and running;
>
>comments of designer and analyst on what, free of restrictions

Conversion to the Final Design, and Implementation 81

or with different restrictions, they would have done differently and why.

Note that, throughout, the report must deal with manual and conventional machine requirements, as well as with computer requirements.

The set of proposals should be considered in detail by everyone concerned with the management of the information function and agreed in principle.

RELATIONSHIP BETWEEN PRELIMINARY SYSTEM AGREEMENT AND FINAL SYSTEM

It is important to note that this Preliminary System Agreement is still only an agreement on the overall design of the system. It is, of necessity, still vague, especially about details of design of the computer system. In particular the file and record designs are there to enable the laymen to appreciate the designers' approach to the problem; the file layouts in particular are not those which will probably be implemented in the end. In the final stages most of the information staff are not – nor do they need to be – responsible for the turning of an idea for a system into a system. Nor, often, are they sufficiently well informed even to conceive of the problems involved. From this point onwards this book is concerned merely to indicate to them what to expect and how to deal with it. Therefore, however 'preliminary' the design may be, it is very nearly final as far as the staff of the information department are concerned. From now on, major changes in either the overall concept or in some apparently minor point would cause trouble.

MANUAL WORK INVOLVED

It may seem strange to write in this book of the non-machine work involved, but there is yet much to be done by the staff of the information department.

In the first place it is desirable (though not essential) to set up formal lines of communication between the systems analyst (who, it will be remembered, is responsible for defining the operations by which the new system will be kept working), the system designer (responsible for working out the techniques by which they shall be implemented), the management (holding final responsibility for the success of the operation and in particular holding the purse strings) and the other information staff (who will eventually make or break the new system according to their understanding of it and their efficiency in operating it). Where possible regular meetings between these four should be arranged throughout the forthcoming stages, even though during these stages the bulk of the work will fall on the shoulders of the system designer. Even at this early stage it is not too soon to begin designing the test material, primarily a job for the information staff and for the systems analyst working on their material. In the first place this work will probably involve simply compiling actual information in the form they think best; later still they will have to see it through the various data-preparation processes (and here, too, modification may well be needed to fit it best for the handling it will receive in these processes). All this time, of course, they will have been causing their own amendments to be made and the information staff must be carefully prepared in advance for the 'interference' of the others in their work and design. It is not too much to expect that over all about 20 per cent of the information staff's time is going to be diverted to this work, though careful planning (e.g. by getting the indexers to set out their work according to new form layouts) can cut this down quite a lot. This is where some new practical ideas may be initiated but they must not be allowed to interfere with the final plan of the system and must always be subject to the consent of the system designer who after all has already started others on their parts of the implementation plan.

This stage also is the latest to start arranging for the costing of the manual part of the work. This part should have been begun as soon as possible – often long before this stage – but it is rarely that the existing costs of the comparable operations have been worked out: overall costs, yes; detailed costs, no. If the best is really to be got out of the system, the cost of every smallest operation must be taken into account. These considerations should have been taken into account in drawing up the 'final' preliminary system; but often the data for them were insufficient and it will be necessary for provision to be made in the system staff's later drafts for greater economies in the manual parts of the operations. Of course, the cost of manual operations will probably increase as the new system is implemented. But there are in the last resort

Conversion to the Final Design, and Implementation 83

only two reasons for mechanizing – to do the job better, or to do it at lower unit cost. The balancing of one against the other is for the senior management and at this stage they will be keeping a careful eye on the cost effectiveness of the proposed changes.

Later, too, the information department staff will be required first to test the outputs from the draft programs, and then to try them out on the clients. Careful selection of such users must be made; there are many who object to the new simply because it is new, just as there are some (though fewer in number) who approve it for no better reason. Arranging for such testing, and for reporting on it, is essential and must be provided for.

Finally, the information department staff have to plan ahead for the physical handling of the output of the computer, and for coping with the new systems and routines for the changed flow of work which will result.

One of the things which the new system will have to accommodate is provision for some small changes in the indexing language. Fortunately this is also the right time for making final tests of that language. Its composition and the principles of its structure should have been worked out long before and it is too much to expect the analyst and designer to include in their programs provision for change from one kind to another of indexing language. But provision should have been made and should now be tested for making minor variations in either the indexing of a single item or in the relationships between two or more indexing terms. If the chosen new system is to include a computer-held indexing vocabulary, provision must also be made for changes in it.

As programming progresses, the information staff can expect to be asked to produce a variety of examples of potential mis-handling on their part: the system designer will almost certainly produce list after list of conditions in which errors might occur; he must build his system in such a way that these can be handled if not automatically at least with the minimum intervention on the part of the information staff.

IMPLEMENTATION: PROGRAMMING AND TESTING

Objectives

The purpose of this stage is to get the program actually running.

The programmers will prepare detailed flow-charts for the intra-computer work; but in addition those of the systems staff must cover the final version of the whole system – human and mechanized operations.

The final descriptions are made (probably during programming) of every input, file and output. The first and last named will also be needed in the *Users' guide*, whose preparation starts during this final stage.

Programming will not necessarily be sequential through the system: it is often helpful to the programmer to work from last to first, because by tackling the job in this way he can spot omissions (from the earlier drafts) of checks and signals needed in the working program and not realized until then.

During programming there will be appreciably fewer discussions between computer and information staff, although some will be inevitable. Each must be recorded in some detail; it will not be sufficient merely to define the agreement arrived at, but to give also the pros and cons of its being made. These will be useful in designing test material and extremely useful when the system is added to or changed in any way.

The complexity of the programming is primarily a function of the 'language' in which the instructions are written; some have instructions almost entirely in some kind of code, and are not easy for the outsider to read and for the programmer to write; others are less complex to write but quite as hard to understand (see Appendix 1). No hard-and-fast rule or even general guidance can here be given as to the time taken to write a program: the information staff must here be guided by their computer staff.

Specialization

In the preceding chapters, an attempt was made to differentiate between the systems staff and the information staff. Often the same people act in both capacities and, especially with the advent of programming languages closer to ordinary speech, it is becoming more common for the information staff to undertake the system design and programming functions; only rarely does the converse apply, yet for information scientists to imagine that they can undertake the work involved in putting an office 'on to the computer' is as foolish as for them to imagine that a programmer will of necessity be also a good cataloguer or indexer. Information staff should beware of trying to do work for which they have not been trained. If they are left to cope with the burden of computerization, then they should bring in professionals wherever possible.

Stages of the Work

The stages through which the work must pass often overlap, and forecasting the manpower requirements is complicated; it should be recognized from the start that the information staff will need to do extra work during the early days of the work on the computer-based system. It must not be forgotten that the computer-based service will not save labour: it will only permit more work to result for the same effort. Management must never underestimate the manpower and time that will be required to keep the new system working. The following is a summary of the main phases through which the work must go; it must be read always with the thought in mind that many of the phases will overlap.

System agreement with the user. The systems staff and the information staff working together must agree what is to be achieved by the proposed system (the statement of goals should have been prepared in advance) and should specify what is done by the present system and where it fails to meet the demands of the organization; this statement should also make quite clear what constraints of finance and/or equipment are to be put on the new system.

Statement of goals. This more specific document should be made by the policy makers responsible for the work which the new system will serve. It must specify the existing scope of the work proposed for computerized operation, in terms of the functions filled and the purposes to which they are directed. The deficiencies of the existing system should be clearly indicated, with proposals for overcoming them, including functions desired but hitherto regarded as impossible of achievement (these should still be included when they are considered impossible even with the new system). The relative priorities (in terms of time and of possible cost) of all the listed goals must be made explicit.

Statement of requirements. This should be made by the systems staff and should specify in detail all demands to be made on the proposed system, the data required for each (what they are and whence they derive), how they are operated upon and what data are produced. (N.B. 'data' is here not confined to its literal sense but is a short word for all information handled by the system.) It should include the proposed inputs, files and output in sufficient detail to clarify the interactions of the various sub-systems that will be required: these sub-systems should be distinguished and described in outline.

Flow-charting. This concludes the genuine creative design work, so all necessary information should have been gathered before now. The flow-charts and their associated documentation should now show what data are being handled at every stage with details of the machine requirements (if any: this work covers extra- as well as intra-computer activities) and the possible costs of each operation.

Preliminary system agreement. This must be between the systems staff and the information staff and management. The feasibility study on which this stage is based must have the whole-hearted assent of all interested parties. The Preliminary System Agreement is a more or less formal agreement on the overall design of the proposed system. The nature of the machine constraints has already been indicated; now they are to be specified.

System design. This takes the proposals in the preliminary agreement and converts those that are to be computer-handled into a computer system. Here the designer is concerned with balancing input and output, with details of file design, error detection and correction, and the relationships between sub-systems. The whole structure of the system is sub-divided into coherent sections and these into individual programs.

Considerations to be taken into account in file design include speed of getting the file into core; the cost of maintaining and using the file; operating difficulties and ease of access to individual items of data; the volumes of information to be held and the effect of these (both on file design and system design); the medium on which the file is to be held. Similar considerations affect input and output design.

Conversion to the Final Design, and Implementation 87

Programming and test data. One result of the programming phase must be yet more documentation, more detailed, of every individual program, and during programming the types of test materials required by the program can be specified and work begun on the preparation of test material. The amount of work required in preparation of test material is often underestimated. During this phase, too, work can be begun on the preparation of the *Users' guide*.

System testing. When the programming has been completed and each individual part tested, the system as a whole must be tested. The users' role in the system must also be exhaustively tested. Work should be completed on the *Users' guide*.

Handover. The information staff are responsible for ensuring that all the errors they can anticipate have been catered for, and it may well be desirable to arrange for the simultaneous running of the new system with the former system, before the latter is abandoned.

Maintenance. A regular (six-month) check should be made on the performance of the new system. Any change in a part of the system must repeat the whole of the above processes: patching in a new part, or deletion of any part, must not be undertaken without the same stringent discipline as was applied in the creation of the new system. There is a real danger of upsetting the entire system by altering only a part of it. Changes may also be required by the advent of new hardware. Every change in the system must be fully tested. Provision must be made therefore for keeping communication going between systems and information staff. Control of the system must be continuous; even so simple a thing as re-design of one form must be carefully monitored.

File security. Once the content of a file has been altered, the alteration lives on, possibly indefinitely. Every precaution must therefore be taken to protect a file from accidental over-writing. Hardware can help a little with this: a so-called read-write ring (see Appendix 2) can be affixed, preventing any access to a file without physical attention having been given to it. Partly it can be protected by program, with a programmed requirement to check that the correct file is being addressed, with a first record (showing the file's creation date, updating date and contents) being maintained by program. Some security can be effected by maintaining successive files intact for a couple of runs, so that back-tracking to an earlier date is possible: if, for instance, there has been an error in handling the tape today, it is a security to have a copy of the tape produced by yesterday's run or even its predecessor (these successive generations of tape are known as 'son, father and grandfather tapes'). Another precaution is to use a 'dump' tape for all material discarded as a result of a run, so that material discarded in error can be retrieved. All these precautions should be taken as a matter of course by the computer staff but the information staff should be continually on the look-out to ensure that the data present on their files are properly protected.

UNDERSTANDING THE COMPUTER

It is clear, therefore, that the appointment of systems staff does not exempt the information staff from continuing concern with the well-being of the computer-based system. Most important, it is necessary to avoid a common misunderstanding of the nature of a computer. It is not an intelligent machine. It appears intelligent in that the program can require certain action to be made if the content of an address is positive, negative or zero, but even these reactions are conditioned by the program. Every circumstance has to be foreseen and answered by the programmer. (This is not to say that the machine never fails, although a computer is a very reliable machine; however, its failures are commonly on a large scale and can immediately be observed.) Mostly, errors committed by the computer are due to programming errors, and errors in understanding how the computer works (that it is limited to simple, one-step-at-a-time actions) lead to misapprehensions about what can be expected of it. Every possible circumstance has to be envisaged and programmed for, and it is here that the information staff have a continuing responsibility to ensure that the systems staff are fully informed about every conceivable contingency. Inadequate anticipation of often quite minor complexities is the commonest of all reasons for the cost of getting on to the computer being so high.

It should also be recognized that a *perfect* solution of the require-

Conversion to the Final Design, and Implementation 89

ments of any system is virtually impossible to achieve; every solution has to be a balance of needs against economy. Continuing discussions between information and computer staff are a prerequisite of achievement of this balance. It is for this reason that the information staff must, quite early in the project, resolve for themselves which elements of the wished-for solution are essential and which merely desirable adjuncts. The computer staff must know the objectives of the information department and they should examine the tools, machine and intellectual, conventional or computer, hardware or software, which could be used.

The information staff have a large and continuing part to play in system design. It is not difficult to say, of an element in the total: 'this can well be left out'; it is difficult to say 'this must be incorporated' when no provision has been made for its incorporation.

Briefly, then, we can say that the implementation of a computer-based information system requires complete understanding of the nature of the information process and of the people and tools which are to be used in the process. Without both, the work is difficult indeed, if not impossible.

APPENDIX I

Elements of programming, or how to make the computer do what you want it to do

WHAT A PROGRAM IS

'The average is the total of all the values divided by the number of values totalled' is a definition. It does not explain how to arrive at an average figure, though the reader can make a very good guess from it. To calculate an average one needs a statement like this: 'Add the first value to the second value; make a note that you have the total of two values. Now add the value of the third to the total, and add one to the figure you have for the number of values added; repeat this till the last value has been added; divide the total by the number of values added; the resulting figure is the average.'

This is much more like a program. A program specifies what steps are to be performed, and in what sequence. It instructs the computer to use devices which are part of its hardware by representing these devices symbolically and sets out the instructions in the sequence which causes the required job to be performed. The computer 'reads' each instruction (in fact, the pulses representing the instruction pass successively through it), interprets each program step in terms of its hardware and the charges stored in its hardware, and obeys it. The program steps are stored sequentially so that each has an address in store, and are followed sequentially unless one of the instructions is to jump over (forwards or backwards) to a specified instruction.

An imaginary programming language

The set of symbols used to represent computer operations is the programming language. The bulk of this appendix is about an imaginary but not untypical language, and towards its end it will mention specifically some real programming languages.

Appendix 1

One of this language's most powerful symbols is ':='. This means 'set the value indicated on the left-hand side to be the value indicated on the right-hand side.' The 'value' can be a real value, or it can be the *address* in which a value is stored. And these addresses (*locations*) can be specified symbolically, too, by giving them names. So:

SUM := SUM + 1

means 'set the contents of the location I have called SUM to hold that already in SUM plus one.' SUM is a symbolic address; := is a symbolic instruction, and + is a symbolic operation. The whole is a program step (loosely called an instruction). To calculate the total frequency of all the frequencies of keywords, could, for instance, be achieved by an instruction:

SUM := FRE(1) + FRE(2) + FRE(3) . . . . . .

For purposes of illustration, suppose a program to be required to calculate the average frequency of the keywords used in a collection of documents. The frequencies have been stored in a series of successive words, the first word of the series holding the frequency of the first keyword, the second holding the frequency of the second keyword, and so on.

To calculate an average one must add the second to the first frequency and count how many values have been added, and so on. So it will be necessary to set up, in store, a location which we shall call COUNT. And the first step is:

COUNT := COUNT + 1

meaning 'set the value in COUNT to be 1 bigger than it presently is.'

Symbolic addresses can be used to specify both locations holding values and locations holding instructions (after all, any computer word, be it value or instruction, is still only a 'word' in store holding a string of 1s or 0s). So let us call this instruction (LOOP) (the reason why will soon become apparent); any names may be used to represent locations – Able, Baker, Charlie, Don; Pig (1), Pig (2), Pig (3) ... FRE is used here as the symbolic name for the frequency of a keyword, and FRE (1), FRE (2) ... FRE(N) for the locations holding the 1st, 2nd ... Nth frequencies. Again for purposes of illustration let us decide that in this case we have 365 keywords, so N = 365.

We shall need a location for holding the sum of the frequencies (SUM), and one for the average (AVG). So now we have 5 symbolic names, SUM COUNT FRE AVG N. Only N holds the same value from beginning to end of the program – it is a *constant*, the others are *variables*.

Now, it is necessary to add the contents of the first frequency to SUM, then of the second and so on. Since they are being handled successively, the value in COUNT can be used to say which particular frequency is being added, and the next instruction can, very simply, read:

SUM := SUM + FRE(COUNT)

(Add to the contents of SUM the contents of the COUNTth frequency; store the result in SUM)

Next it must be decided whether the computer has come to the end or if there are more additions to do. Once the value in COUNT reaches 365 (a constant which we are calling N) we shall have added the last frequency to SUM, and can go on to calculate the average. If COUNT holds less than 365, we must go round the loop again; so

IF COUNT < N GO TO (LOOP)

Of course, if COUNT contained anything left from a previous program we would be in trouble. So we must *initialize* by setting COUNT and SUM to ZERO, and setting N to 365. (Common computer jargon is to 'initialize' [have to set, at the start] COUNT = 0.)

Now that 365 frequencies have been totalled an average can be computed

$$AVG := \frac{SUM}{COUNT}$$

The computing part of the job is done. But it is still necessary to print the results out. A heading and date are needed (*DATE* is normally held in store throughout the day in any computer, and can be addressed by any program), and they are followed by column headings defining the data. The program language has an instruction *WRITE* which operates the line printer, and following *WRITE* the computer is told what is to be written. But we are up against

Appendix 1 93

trouble here, for we have used English language words as symbolic locations and instructions in our program, and now want to use them as real English language words in our print-out. The program language gets over this by distinguishing: symbolic words are not put in quotes, words to be printed out as they stand are put in quotes. When we want to start a new line, we finish the *WRITE* instruction with a * for every line thrown and '∗' represents a space. So we get the following bit of program:

WRITE 'KEYWORD FREQUENCY COUNT ∗ ∗ ∗ ∗ ∗'
         DATE*
WRITE 'TOTAL FREQUENCIES := ∗' SUM*
WRITE 'NUMBER OF KEYWORDS : - ∗ 'COUNT*
WRITE 'AVERAGE FREQUENCY : - ∗' AVG*

and that is the end of the program so:

STOP

Parameterizing

This program is limited. It *only works* if we have 365 keywords. To be much more useful, we should write it in such a way that the constant N can be set to a pre-defined value. This is called *parameterizing*: N is a parameter which, constant for any one run, can be varied at will at the beginning of any one run. Let us go back to the beginning and drop our assumption that the keywords are already in store. We will read them in from a card reader, and precede them by a parameter card holding the value of N.

Each card holds the value at the beginning and the READ instruction asks us to specify from what card reader, how many cards, into what locations:

READ CRI   I   N
READ CRI   N   FRE(COUNT)

This means read one card from card reader I and put its contents into the location called N. Read N cards and put their contents into locations called FRE, any individual one of them being called FRE(COUNT).

This complete set of instructions must be identified within

the computer *as* a set of instructions, not a set of records, or values or anything else. Without such an identification, it is nothing but a string of os and 1s. So it is necessary to prefix any program with ‡ ‡ = PROGRAM and the name of the program: KFCOUNT and then all we need to do is the initial instruction *START*.

Obviously it is not as easy as all that. Records do not comprise one value each; and in any case the program has to be instructed about the number and kind of every record, and the relationships between the records, in a real situation. It has to be prefaced, too, by details of the way each record will be addressed symbolically. But this gives the general idea.

This program was parameterized on only one item – the number of records. But programs can be made increasingly general-purpose by providing sub-routines which can be called into the main program by a *JUMP* instruction. For instance, a program may allow paper tape OR card input: two separate *READ* programs will be needed, one for each, but only the one required by the read parameter card will be used. There are two things to note about this imaginary programming language. The first is that English language words, plus a few symbols, have been used as:

(i) instructions (I̲F̲, *GO TO* – distinguished by being in italic capitals or underlined);
(ii) literals (words to be treated just as they stand – distinguished by being in quotes);
(iii) modifiers (which alter addresses or instructions and are given in brackets);
(iv) addresses of instructions (a form of modification) (given in brackets, e.g. (*LOOP*));
(v) instructions such as :=;
(vi) operations such as +, −, /, etc.

The second thing to note is that this program is very small in spite of its having 365 separate additions of frequencies; add-1-to-count and test.

This is because of the power of the computer to jump to various parts of the program according to the result of test operations. One of the problems of writing large complex programs is that one

Appendix 1

cannot always be sure that one has allowed for every alternative, and that one has allowed for it correctly. (For example, if in this program, one had put SUM := SUM + FRE(COUNT) before increasing COUNT there would have been trouble in coming out of the loop. And programs can loop within loops, and from within a loop they can jump to a loop within quite a different loop and so on, so that if the programmer is not careful, he might have a program which would go round and round without ever coming out of the looping.)

MACHINE-CODE PROGRAMS

A computer instruction actually takes a rigid format; the steps written above have to be translated into sets of binary words before they can be used. In the early days of programming the only symbolism possible was to represent a basic machine instruction in denary instead of binary (see Appendix 2). Eight bits were allowed for the actual operation code, but we were allowed to use 001 for *ADD*, 002 for *SUBTRACT*, etc. The actual address of any 'word' in store had to be used, not a symbolic name for it. The format went like this:

Instruction no.	Operation	Accumulator	Operand	
001	003	2	3927	Move contents of 3927 into acc. 2
002	001	2	5628	Add contents of 5628 into acc. 2
003	003	2	3	Move contents of 3 into acc. 2
004	002	3	9305	Subtract contents of 9305 from acc. 3
005	011	3	010	Jump to 010 if result (in acc. 3) zero
006	003	2	001	Move inst. 1 into acc. 2

Instruction no.	Operation	Accumulator	Operand	
007	001	2	6000	(Location 6000 holds 1 and this instruction adds 1 to address in inst. 1, now in acc. 2)
008	003	2	001	Move inst. 1 into acc. 2
009	010	0	001	Jump to inst. 1

Instructions were modified by reading them into the accumulator and adding a constant. The *JUMP* in instruction 005 could not be written until the computation of the modification of the operand in the beginning of the loop had been written. One had to remember where everything was stored, and have a number of constants for modifying instructions with, and programs in consequence became enormously long and difficult to keep track of even while writing them.

PROGRAMMING LANGUAGES

All actual machine instructions, even to this day, are of this form – more or less, according to the hardware – except that they are in binary. Each instruction can do only one simple thing, and modification was an appalling headache, as was keeping track of the use of store. *Programming languages* free the programmer from this detail. In the very (so-called) low-level languages, there is a nearly one-to-one correspondence between the programming language instruction and the machine code instructions generated from it; the computer needs only an *assembler* program to do the conversion from the *source language* as written by the programmer to *object language* in machine-code. Take PLAN, for instance, the ICL 1900 programming language. A PLAN instruction is as formally organized as my machine-code instruction format. There are specified fields, of specified length, for the label, the operation code, the accumulator to be used, and the operand. The programmer does not have to number each step, but gives a symbolic name in the label field for addresses to which the program must jump: (c.f. (LOOP) in the little program above). The operation is written in a (more or less) mnemonic code of three characters. The actual accumulator must be identified, but a symbolic address can be used for the operand. So there is not a lot of work for the assembler to do in converting from source to object program.

High-level languages

On the other hand, a high-level language generates whole chunks of program steps, including the test-and-jump instructions. For instance, after setting the value of N in our example, the program could be:

DO X = 1 (1) N
SUM := SUM + FRE(X)

meaning add FRE to SUM for values of X ranging from 1 to N, by steps of 1, i.e. by making X equal successively 1, 2, 3, 4 ... The object program will be just the same as before, the test-and-jump conditions being deduced and programmed by the computer. (Thus, if the program was 1 (10) N it would have been every 10th frequency which was added to SUM.)

This is a *high-level language*. High-level languages are not so easy to translate into machine-code, and commonly they go through two translation programs – one to turn them from high- to low-level, which is the *compiler program*; and one, the *assembly program*, to turn them from low-level to machine-code, i.e. object program.

High-level languages are user-oriented (which accounts for the OL – Oriented Language – at the end of many of their acronyms). But one which has come to the fore is FORTRAN (FORmula TRANslation) – probably because it was amongst the earliest, and has a limited-character set available on nearly all computers. Unfortunately, it developed a series of 'dialects', each dialect requiring its own, slightly different, compiler.

The other principal scientific language is ALGOL which is much more rigidly controlled than FORTRAN and produced virtually no dialects, so that an ALGOL compiler for one machine will do just as well for any program written in ALGOL for another. It is based on a very wide character set which permits many of its expressions to correspond almost exactly to mathematical notations, but, for that very reason, does not lend itself readily to computers with only a limited character set.

COBOL (Common Business Oriented Language) is easy to write: it is a high-level language using nearly normal English-language expressions. However, I myself find it less easy, since the rules which it makes regarding one's definition of file structure and peripherals (which, of course, must be defined, in every programming language, as part of the programming) are rather forbidding. COBOL has developed an enormous number of dialects, each, of course, needing its own compiler.

There are many other high-level languages, some much more useful (potentially) for non-numeric processing, being designed for list-processing (see Chapter 3, pp. 31-5) or for handling strings of characters (string processing languages). Most of them are still one-off jobs, being written to do one job on one machine with a compiler only for that machine.

Low-level languages

A low-level language is nearly always suited to one breed of computer only, because of its nearly one-to-one correspondence with that computer's own machine code. The higher languages can be used on any breed of computer which has, in its own object language, a program to compile source to object program. Thus, if you had a program, written in PL1 for an IBM 360, and you only had a 1900, you could not use it, however suitable it seemed; nor could you use a PLAN program if you only had a 360. But a program written in FORTRAN IV (provided you have a FORTRAN IV compiler in your program library) can be used on almost any machine – within limits, of course. If the program called for a store of 32,000 words (a '32K' store) and disc, and you have a store of 8,000 words and tape only, then you still could not use that program. If you want to use a program not written specifically for you, you must be sure (a) that you have a compiler for not merely that language, but that dialect of that language, and (b) that you have the right configuration (sort) of hardware.

READY-WRITTEN PROGRAMS

These are very numerous. There are the applications packages

Appendix 1   99

(e.g. programs to do KWIC, or handle connection tables); these will be either in high-level language or already compiled (if they are part of the manufacturer's software).

Then there are the programs that the manufacturer will be supplying almost as a matter of course (though there is a growing move towards charging for these programs). These do the standard things like sorting and merging, or handling the commoner mathematical operations; they are not programs which do a complete application, but are likely to be called in as one or more parts of a variety of applications. There are the programs which handle the so-called housekeeping: transferring data from tape to core, or core to disc, etc. These are the *operating systems* and are almost as much a part of the computer as its hardware.

Standard software is normally provided in semi-compiled form which leaves the user free to specify his parameters. The 'housekeeping' software is normally written in object language from the start. 'Housekeeping' is the routine work of keeping programs and internally-stored program libraries in proper functioning condition.

Standard software can either be called during the compilation run of a programe written in a high-level language, and inserted into the object program; or it can be called as a separate run during a suite of program runs.

MULTI-PROGRAMMING, REAL-TIME OPERATION

In the majority of programs except the real 'number-crunchers', the core is only being used for a fraction of the time of the program's run. Most of its time is spent waiting for information to be read from or written to a peripheral or backing store. The obvious thing to do, therefore, is to run programs in tandem: while one is waiting, the other can be getting on with its computation, and then when program no. 2 has to wait, program no. 1 can start up on its newly read (or written, as the case may be) stage. Nor need this be confined to just two programs: a series of programs can be run, apparently simultaneously, but in practice each doing a separate operation: a bit of computing, or reading, or writing. The only thing that cannot be done virtually simultaneously of course, is actually to compute – to run an instruction through the mill, interpret it and act on it. But while program A is waiting many microseconds for a card read, and program B is waiting a few milliseconds for a tape write, and C is waiting several nanoseconds for a disc read, program D can get on with doing some of its instructions.

Of course, a program is needed to control all this; the *executive program*. It is held permanently in store and looks after the sequencing of the switching between programs, optimizing the waiting time for each, giving priority to programs of importance (if such importance has been notified to it), putting into backing store a program temporarily suspended while waiting for a peripheral unit (such as a disc-storage device) to become available (and, of course, noting just where that program was, so that when the time came it could be re-started without any difficulty). Executive programs distribute the core between the various jobs on hand and may vary the core addresses during the run. However, this is not a thing you will ever have to worry about. All you need to know is that several programs can, with a multi-programming facility, be run apparently simultaneously.

A real-time system is one which responds to you, the user, whenever you want, via your terminal. So, for the entire time, there must be in core at least enough program to recognize that you are calling and to find out what you are after. It can then call down from backing store the program to do the job you want and the data on which you will want to operate.

Perhaps you will not want either to be called – you may be using the machine to do some calculations for you and you want to enter program and data as soon as you have attracted the machine's attention. But for an on-line ISR job, the program is quite large, and the data base is very large, and they must both be available – though not necessarily in core – during the entire time you may contact the computer. It would be no good having the base on a tape, which has to be specially loaded each time it is used!

You can have real-time without multi-programming, but if the usage is not very heavy, this can be very expensive. While you are looking at the displayed thesaurus and deciding what term to use, the computer could get a large part of the pay-roll computed, or be doing the searching for another enquirer at another terminal.

APPENDIX 2

Glossary

Access (*Noun*) The operation of accessing. (*Verb*) To get at a part of a computer-held chunk of information.

Address (*Noun*) The location at which a piece of information is held in the computer store. (*Verb*) To instruct the computer (by program) to go to an address in a storage device – in core, on disc, etc.

Algorithm (*Noun*) A series of program instructions performing a specific procedure; thus a search algorithm is the procedure, and the operations effecting it, of performing a search.

Assemble (*Verb*) To convert a source program into an object program.

Assembler (*Noun*) A program which assembles another program. To say that a machine can handle X language normally means that it has an assembler to assemble X language into its own internal language.

Binary (*Adjective, often used as a noun*) A system of numerical representation using only two digit symbols, normally the numbers 0 and 1. Each position in the notation means a power of two (c.f. denary, used for arithmetic in most countries, in which the position of the digit indicates a power of ten). In binary the positions, reading from right to left, mean 1, 2, 4, 8, 16, 32, 64, 128 ... and so on. Thus the binary number 10100 represents the denary number 20 being, reading from left to right, $16 + 0 + 4 + 0 + 0$. In denary, since each position can represent one of ten values, the figure 20 means $(0 \times 1) + (2 \times 10)$. Binary is used in computer working since the electrical circuits used as a means of representing characters (letters or digits) have only two states – positive or negative; so a positive charge can represent a 1 and a negative charge a zero.

Bit (*Noun*) Now a word in its own right, this was originally an abbreviation of 'binary digit'; thus the binary number 10100 (value 20 in denary) is made up of five bits. The expression 'bit position' is commonly used; to represent denary 4, for instance, requires a 1-bit in the third bit position.

Byte (*Noun*) A group of bits handled by some types of computer as the normal means of representing characters. A byte may comprise, for instance, sixteen bits. (*See also* CHARACTER and WORD.) Many computers measure store etc. capacity in Kilobytes (KB). 1 KB = 100 bytes.

Card, punched (*Noun*) A card reader senses the holes present in a punched card and 'reads' them, normally into the core in the first place. The positions on a card are arranged in (normally) 12 rows, each with 12 punching positions numbered 0 to 9. 'Reading' is effected by passing the card between 80 sensors and a metallic roller: where there is a hole the sensor makes a connection with the roller and a current can pass through. The position of the hole thus sensed indicates its value. There are thus 80×12 positions on each card and normally only one denary (*q.v.*) value is represented in each column but sometimes each of the 960 positions has an individual significance, the presence or absence of a hole in each representing either a 0 or a 1 value for that location. Alphabetic characters are normally punched one to a column, each character being represented by two holes in that column.

Reading speed is normally expressed as cards per minute and in modern machines varies from 300 to 2000 c.p.m.

A card punch operates similarly, translating characters in core into the appropriate punching positions in a card, and causing the corresponding holes to be punched. Punching rate normally 100 c.p.m.

Character (*Noun*) A group of bits representing a symbol – commonly but not invariably an alphabetic character. The number of bits allocated to any character position varies from one make of computer to another: alphabetic characters commonly take up five or six bits each. Often two different configurations of bits represent one the lower case and the other the upper case of an alphabetic symbol.

Cluster (*Verb*) A mathematical process in which the characteristics of a number of discrete objects are analysed and the objects brought into clusters, the members of one cluster being more like each other than they are like the members of any other cluster.

Constant (*Noun*) A value held unchanged during the course of computation. Thus in the expression πr^2, π is a constant, as opposed to r which in this computation is a variable.

Core (*Noun*) Properly 'core' (or 'main') store—part of a computer in which all processing is carried out because all components of the core store are equally and very quickly accessible. (In the early days of computing the core store was generally known as the IAS – Immediate Access Store – because each element in it could be so quickly accessed.) Manipulation (e.g. the adding together of two numbers) is done in a special part of the core known as the Arithmetic Unit; the results may stay in core awaiting further processing or they may be 'written' back into one of the storage devices (*see* STORAGE DEVICE). All reading from an input device puts the newly-read material into core, whence its writing to store is directed by program. Similarly all writing to an output device takes place from core, into which the output material has first to be read from one of the storage devices.

Denary (*Adjective, often used as a noun*) A system of arithmetic notation in which the location of a digit indicates the power of ten over the digit to the right of it (*see also* BINARY).

Descriptor (*Noun*) Anything which describes an attribute of a subject. It includes alphabetical subject headings, class-marks or numbers, keywords; it also includes the codes used to represent these things.

Disc (*Noun*) A storage device which can be likened to a gramophone record: data are stored on it in tracks. Here the resemblance breaks down, for there is not a single continuous spiral track but a series of discrete concentric rings. Addressing a disc comprises in fact two operations; selecting the appropriate ring by moving the read–write head in or out, and selecting the correct part of the ring (the sector). Once the head has got to the desired ring, it has to wait until the continuous revolving of the disc brings the required sector under the head. Discs can be fixed or exchangeable (*see also* CORE, DRUM, STORAGE DEVICE).

Drum (*Noun*) A storage device not unlike a disc save that the rings of recorded data are recorded in tracks running round the outside of a drum. Access is in two stages, as with a disc; moving the head to the correct ring, and then waiting till the required part of that ring comes under the head. Drums can be very large, holding vast quantities of data; being of uniformly large circumference, with the drum moving more slowly than a disc, access to any one part can take a (relatively) long time (*see also* CORE, DISC, STORAGE DEVICE).

Dual dictionary (*Noun*) A printed inverted file, comprising two identical parts normally printed side by side, with one binding for the two, so that a (non-mechanical) comparison can be made of the document numbers under each descriptor. To facilitate this comparison, the document numbers are frequently set out in ten columns, arranged so that the last digit of each number is the same in any column; within the columns the numbers are in ascending order.

Feature card (*Noun*) A form of inverted file, designed to facilitate non-mechanical scanning. Each card contains the numbers proper to the descriptor which the card represents; printed on each card there are numbered tiny squares, each representing the document having that number. If the descriptor applies to any document, a hole is punched in its square. When the cards representing, say, descriptors A, B and C are superimposed, there will be a hole right through all three for any document indexed by all three descriptors.

Field (*Noun*) A separately identifiable section of a record (*q.v.*) in which a piece of information is stored. Thus one can have a field for date, a field for title, a field for author. Fields are known as *fixed* or *variable*; a fixed field is of the same length in every record of one kind (thus 'date of birth' could be a fixed field in a file of personnel records) whereas a variable field is not; thus a file of personnel records might have a variable length field for the name of the individual, which might be Smith or Montgomery.

Not to be confused with fixed or variable, which refer to length, are *variant* and *invariant* fields. A variant field is one which at different times (either during the life of the record or during one processing of the record) holds different values, as opposed to an invariant field which holds the same value.

A *repeating* field is one which occurs more than once in the same record; thus, if there were six descriptors to a document record, the descriptor field would occur six times, with different contents in each.

Graph plotter (*Noun*) An output device in which the output data are printed as points on a graph. Such a device is useful where such things as frequencies are to be represented, but is is not commonly used in ISR.

Housekeeping (*Noun, sometimes used adjectivally* as in 'a housekeeping program') The routine maintenance of programs and the contents of the computer. Most housekeeping is undertaken by a program or programs supplied by the manufacturer of the computer. The staff of the Information Department, even if they are doing their own systems and programming work, need have no concern with housekeeping programs.

Initialize (*Verb*) To prepare by program; to set (usually the contents of a store, or a constant) to a suitable initial state for the required processing.

Instruction (*Noun*) A single program step. The word, originally meaning a single machine instruction specifying what kind of operation (*ADD, MOVE, WRITE*, etc.) is to be performed on what operands and where the result is to be placed, has come to mean any single program step in the programming language. Such a step, if the language is 'high-level' may in fact involve several computer operations (*see also* LANGUAGE).

K An abbreviation for 1000, used for example in Kg for kilogram. A 32K store is a store capable of holding 32,000 words.

KB *See* BYTE.

Key (*Noun*) That part of a record which identifies it for the purpose of an operation to be performed on the record. Thus, if a list of author names is to be sorted into alphabetic sequence, the name is the key upon which the sorting takes place. (*Verb*) To operate the keys of a typewriter-like device, either for inputting or outputting characters to the computer (*see also* KEYBOARD TERMINAL).

Keyboard terminal (*Noun*) A machine resembling a typewriter. Used as an input device, pressure of a key generates a pulse received in core as the pattern of charges representing the character thus typed. Conversely, the keyboard can be used as an output device, translating pulses from the core into pulses which operate the appropriate keys. A keyboard is the commonest means of conversing with the computer: often the program is written to instruct the user what to key in, in response to a question which the program causes to be keyed out. Keyboard terminals first made on-line systems suitable for ISR work. Relative to, for instance, output to a line printer, the keyboard is a slow input/output medium, operating at speeds ranging from 20 to 50 c.p.s. The fast Qume and Diabollo printers, operating on a slightly different principle, have output indistinguishable from that of a high quality electric typewriter.

Keyword (*Noun*) A word or short phrase based on post-coordinate indexing and normally assigned by a human indexer. Some authors distinguish between 'keyword' thus humanly assigned and 'title keyword' or 'text keyword' selected by some automatic means from title or text.

Language (*Noun*) Often used for 'program (programming) language'. The collection of expressions used by the programmer and 'understood' by the computer to represent operations. A language can be low-level in that it is very similar to or the same as the language used by the computer, in which case there is nearly or exactly a one-to-one correspondence between the operation instructions written by the programmer and those used within the machine; or it can be high-level (often very similar to spoken language) in which case one expression in the high-level language represents, perhaps, several machine instructions.

Leader (*Noun*) The first record in a file, or field in a record, which carries not the information normally held in the records of that file, or fields of that record, but information identifying that file or record and its contents. The leader record, for instance, often carries among other things an identification of the generation (1st, 2nd, 3rd, updating) of the file.

Line printer (*Noun*) An output device operating, from pulses output from core, the printing of characters on a sheet of paper. A line printer normally comprises one full set of characters for each print position (across the page) all the 'A's, for instance, coming up to the print position at once and being printed simultaneously. Thus the whole line has to be assembled for printing before the first character of that line can be printed; the characters for that line are usually placed in a 'buffer' store but this operation is usually done by stored program or by hardware. The length of line which can be printed depends on the line printer used. Printing speed, measured in lines per minute, ranges from 200 to 1000. The very fast line printers are known as high speed printers.

Magnetic tape *see* TAPE.

Main store (*Noun*) *See* CORE Main stores are becoming more common not on magnetic core but on semi-conductor principle.

Object language (*Noun*) The language used by the machine itself, i.e. the codes used for operations and operands, in instructions operated upon by the computer.

Optical coincidence card *see* FEATURE CARD

Paper tape (*Noun*) Input/output device, the paper tape reader and paper tape punch handle punch rolls of narrow paper on which characters are punched in a similar way to those punched on card. Being narrow, the tape can accommodate, in one line across it, only a few positions and the encoding of characters in terms of these positions is, naturally, different from the encoding of the punch positions on cards (*q.v.*). Paper tape does not have the limitation of cards to 80 columns; a paper tape record can be as long as required. Often paper tape is generated from cards for input of data. It is read serially, so some pre-sorting of the material is often performed. Paper tape is often used as a medium for transporting data from one geographic location to another. Reading (input) speed varies from 300 to 1500 characters per minute; output (writing) speed is 110 c.p.m.

Parameter (*Noun*, hence *verb* **Parameterize**) A parameter is one of the constraints upon the performance of an operation. One can say, for example, that an operation is to be done for all values between 5 and 100. In this case the parameters are 5 and 100 and the program is parameterized to repeat the operation between the values stated. A parameter can be set initially to a given value (*see* INITIALIZE) and subsequently modified by program.

Peripheral (*Adjective used as a noun*) A generic name for any part of the computer not central to its operation, other than its arithmetic units and its core store. The various storage, reading, output and transmission devices are all peripherals.

Precision (*Noun*) Precision is defined as the proportion of retrieved material relevant to the enquiry. It is a measure of the degree of accuracy of a retrieval process. Normally precision and its related recall (*q.v.*) are dependant on the efficiency not of the retrieval program, but of the indexing language used by the system.

Random access (*Noun phrase*) Access (*q.v.*) to parts of a file in no particular sequence. For instance, if a file holds records serially in alphabetical order, to go straight to a record beginning with N and thence to one beginning with F is to make two random accesses. Serial

Appendix 2 107

access would first run through A, B, C ... N, and then A, B, C ... F, with a potential waste of time. Random access is facilitated by hardware such as disc or drum (*qq.v.*) where any one part of the storage is neither more nor less remote from reading than another. It also demands the maintenance of some kind of record of where in a particular file the sought information may be found. On-line information retrieval normally demands some form of random access.

Read (*Verb*) To scan and interpret, by the computer, information presented to it in any way – on a punched card, on paper tape, on magnetic tape, on a drum or disc, etc. In computer jargon the verb is not restricted to detection of visible signals: audible and other signals are growing more acceptable as means of inputting information to a computer.

Read–write ring (*Noun phrase*) A piece of hardware (sometimes, though rarely, of software) not under the programmer's control but set by the operator, which inhibits the computer's reading from or writing to the devices to which it is attached. It protects the data carried on the device from accidental reading or over-writing.

Recall (*Noun*) Recall is defined as the proportion of relevant matter retrieved from the total relevant matter in the data-base. It is a measure of the degree of efficiency of a retrieval process. Normally precision (*q.v.*) and recall are dependent not on the efficiency of the retrieval program but on the indexing language used by the system.

Record (*Noun*) A record is a collection of data relating to the single entity which is the subject of the record. A record is made up of several fields (*q.v.*) and several records make up a file. (*Verb*) In computer jargon, to record item(s) of information on some computer storage medium.

Reformat (*Verb*) To rearrange, normally the content of a record. For instance, when a record is subject to several different operations, it may be necessary to reformat it for some of them so that they can be performed more efficiently. A record often has to be reformatted for printing.

Relevance (*Noun*) Formerly used for what is now commonly referred to as precision (*q.v.*)

Sector (*Noun*) *See* DISC.

Sentinel (*Noun*) One of a set of easily recognizable signals spread along a file, which being more easily recognized and sought than the individual records, permit quick computer search for the desired part of a file.

Source language (*Noun*) The computer language used by the programmer and translated by the computer, using an assembler, into the object language (*see* ASSEMBLE, LANGUAGE and OBJECT LANGUAGE).

Storage device (*Noun phrase*) A device for storing (as opposed to operating upon) data. Data can be read from the device into core, operated upon there, and written back to the same or to another device, but cannot be operated upon while actually in (or 'on') the device. Many storage devices have the part which actually carries the data detachable (e.g. tape or exchangeable disc store) but some such as a drum are an integral part of the computer (*see also* DISC, DRUM, TAPE).

Store (*Noun*) Any part of the computer in which data can be stored. 'Store' is often colloquially used for Core (*q.v.*). 'It is read into store' is often used when more correctly the speaker means 'it is read into core'.

Tape (*Noun*) A storage medium in which the data is carried by successive magnetic charges (magnetic tape) or by holes (paper tape) which, encoded, represent the data. Tape, usually magnetic, is the common way of storing data outside the computer, both programs and successive generations of records being so stored, and most installations have a large library of tapes. Tapes are transportable, and are commonly used as a means of data transmission from place to place (*see also* PAPER TAPE).

Variable, Variant *See* FIELD.

Visual display unit (*Noun*) An output device resembling a television tube in which the characters read from store are translated into points of light on the screen. A V.D.U. can, of course, output anything which can be represented pictorially – words, numbers, graphs – but its disadvantage is that unless programmed for output on to another medium, the output is ephemeral.

Word (*Noun*) The smallest separately addressable unit of many computers. In some makes of computer the smallest such unit is known as a byte (*q.v.*). A word is comprised as a number of bits (*q.v.*) variously used to encode denary numbers (usually four bits per digit), alphabetic characters (usually six bits per character) and other symbols. These numbers and characters (and indeed, in most cases, individual bits) can be separately addressed in core but not within the word in the backing store. The size of the word (i.e. the number of bits it comprises) varies from one make of computer to another. One speaks of, for instance, a 32-bit word, meaning a word comprising 32 bits.

APPENDIX 3

Speeds of some peripherals

Card: read 300–2000, punch 100 cards per minute.
Paper tape: read 300–1500, punch 110 characters per minute.
Magnetic tape: read and write 20–320 KB per second.
Exchangeable disc: read and write 40–806 KB per second.
Fixed disc, and drum: read and write 875–2800 KB per second.
Line printer: prints 200–1000 lines per minute.
Typewriter (or Diabollo or Qume): write 30–50 characters per second.

(The data for card, tape, disc and drum were provided by the Education and Training Division, International Computers Ltd., to whom my thanks are due.)

Reading List

Because this book sets out to be instructional rather than a gathering together of experience of many people, this is a reading list rather than a bibliography supporting the arguments tendered; and because the division is into chapters purely on the lines of my personal approach to system analysis, very few indeed of the works cited fall uniquely under a single chapter heading. Finally, once one moves away from the strictly practical, the literature on mechanized ISR is legion, and an all-inclusive list of literature on the subject is beyond the bounds of possibility.

I have therefore confined myself to those books and papers which I have found most useful (and seminal: many of them started me thinking on lines till then wholly novel to me) in developing my career of putting ISR systems on the computer, and for each entry, to facilitate scanning, there is given a one- or two-word description of its coverage.

Marked 'Essential reading' are books which ought to have been read by anyone embarking on the mechanization of his information system; and most of the cited works also carry exhaustive bibliographies for those wishing to read further.

ALEXANDER, T. The wild birds find a corporate base. *Fortune*, August, 1964. 130–134.
    Group discussions

AMEY, G. X. The critical features of interactive query systems. *In:* Gamache, A. *and* Peneer, R., *comp.* 2nd Open Conference on Information Science in Canada. Proceedings p. 49–58. Winnipeg: Canadian Association for Information Science, 1974. 301 p.
    Theory

ARNOLD, R. R. and others. Introduction to data processing. New York, London: Wiley, 1966. 344 p.
    Data processing

ASHFORD, J. Software cost: making or buying it. Program, **10**(1). January, 1976. 1–6.
    Costs

BATTEN, W. The mechanization of chemical documentation. *Journal of Documentation*, 32(3). September, 1976. 207–243.
General

BECKER, J. and HAYES, R. M. Information storage and retrieval tools, elements, theories. London: Wiley, 1963. 459 p.
IR theory and practice

BELL, C. L. M. and JONES, K. P. A minicomputer retrieval system with automatic root finding and roling facilities. Program, 10(1). January, 1976. 14–27.
Handling indexing

BOLLES, S. W. The use of flow-charts in the analysis of library operations. *Special Libraries*, 58(2), February, 1967. 95–98.
Flow-charting

BORKO, H. The conceptual foundations of information systems. System Development Corporation Paper No. SP–2057. Santa Monica, California: the Corp., 1965. AD615718
Also in: Montgomery, E. D., ed. The foundations of access to knowledge: a symposium. Syracuse University, 1968. Syracuse: the University Press, 1968. 61–68.
Theory

BOURNE, C. P. Methods in information handling. New York, London: Wiley, 1964. 241 p.
Methods

BOUMAN, H. Optimising information retrieval. *Révue Internationale de Documentation*, 32(2). May, 1965. 46–53.
Theory

BROPHY, P. Cobol programming: an introduction for librarians. London: Bingley, 1976. 152 p.
Programming

BUCKLEY, J. S. Planning for effective use of on-line systems. *Journal of Chemical Information and Computer Sciences*, 15(3). August 1975. 161–164.
Examples; cost data

BURKHALTER, B. R., ed. Case studies in systems analysis in a university library. Metuchen: Scarecrow Press, 1968. 186 p.
System analysis

BUTTERLY, E. Improving SDI search profiles. *Information Processing and Management*, 11(8/12). 1975. 189–200.
Profiling

CHAPMAN, E. A. *and others*. Library systems analysis guidelines. New York: Interscience; Chichester: Wiley, 1970. 126 p.
System analysis (for libraries rather than in IR)

CHAMIS, A. Y. The design of information systems; the use of systems analysis. *Special Libraries*, **60**(1). January, 1969. 21–31.
System analysis and design

CLINIC ON LIBRARY APPLICATIONS OF DATA PROCESSING PROCEEDINGS. Urbana-Champaign: University of Illinois, Graduate School of Library Science. Several annual issues.
General

DOYLE, L. B. Information retrieval and processing. Los Angeles: Melville, 1975. 425 p. ISBN 0 471 22151 1.
General

FLOOD, M. M. The systems approach to library planning. *Library Quarterly*, **34**(4). October, 1964. 326–338.
System analysis in libraries

GARVIN, P. L., *ed*. Natural language and the computer. New York, London: McGraw-Hill, 1963. 414 p.
Theory and practice

GOOM, H. H. A computer-based current awareness system producing both SDI output and conventional abstract bulletin. *Aslib Proceedings*, **26**(3). March, 1974. 98–108.
Current awareness systems

GREENWOOD, K. W. jr. EDP: the feasibility study. Washington: Systems and Procedures Organization, 19.
Feasibility study

HARVEY, J., *ed*. Data processing in public and university libraries. Washington: Spartan Books; London: Macmillan, 1969. 159 p.
All aspects – library rather than I S R

HINES, T. C. Minimizing input effort for computer-based information systems: a case study approach. *Special Libraries*, **67** (8). August, 1970. 377–381.
System design

JESTES, E. C. An example of systems analysis: locating a book in a reference room. *Special Libraries*, **59**(9). November, 1968. 722–728.
System analysis

JONKER, F. Indexing theory, indexing methods and search devices. New York, London: Scarecrow Press, 1964. 130 p.
General

KALLENBACH, P. A. Introduction to data transmission for information retrieval. *Information Processing and Management*, 11(5/7). 1975. 137–145.
    Data transmission

KENT, A., ed. Library planning for automation, based on the proceedings of a conference held at the University of Pittsburgh, June 2–3, 1964. Washington: Spartan Books; London: Macmillan, 1965. 204 p.
    General

KENT, A. Textbook of mechanized information retrieval. New York, London: Interscience, 1966. 2nd ed. 280 p.
    General

KENT, A. and others, eds. Electronic handling of information: testing and evaluation. Washington: Thompson; London: Academic Press, 1967 (i.e. 1968). 319 p.
    General

KOCHEN, M., comp. The growth of knowledge: readings on organization and retrieval of information. New York, London: Wiley, 1967. 400 p.
    General

LANCASTER, F. W. and FAYEN, E. G. Information retrieval on-line. Los Angeles: Melville, 1973. 611 p. ISBN 0 471 51235 4.
    On-line ISR

LEIMKUHLER, F. F. Systems analysis in university libraries. *College and Research Libraries*, 27(1). January, 1966. 13–18.
    System analysis

LINN, P. M. Automated library processes and interdisciplinary information studies. Part 3. Interdisciplinary information studies. Report LUT/LIB/R1. Final Report 1971–1975. Report on Project S/20/40. Loughborough: University of Technology, 1975. British Library Research and Development Report 5252.

MALAGODI, A. Locating the information. *Data and Control*, 2(6). June, 1964. 28–31.
    Especially good on document number coding and addressing

MEADOW, C. T. The analysis of information systems: a programmer's introduction to information retrieval. Los Angeles: Melville; Chichester: Wiley, 1967. 2nd ed. 420 p. ISBN 0 471 59002 9.
    Essential reading

MONTGOMERY, K. L. Document retrieval systems; factors affecting search time. New York: Dekker, 1975. 152 p. ISBN 0 8247 6195 2.
    Search strategies in various circumstances

PATTEN, M. N. Experience with an in-house, mechanized information system. *Aslib Proceedings*, **26**(5). May, 1974. 189–209.
Practical

PRATT, G., ed. Data bases in Europe; a directory to machine-readable data bases and data banks in Europe. London: Aslib, 1975. 71 p.
Machine-readable sources

RICHMOND, P. A. Systems evaluation by comparison testing. *College and Research Libraries*, **27**(1). January, 1966. 23–30.
Evaluation

ROBERTSON, S. E. Explicit and implicit variables in information retrieval systems. *Journal of the American Society for Information Science*, **26**(4). July–August, 1975. 214–222.
General

ROLLING, L. N. A computer-aided information service for nuclear science and technology. *Journal of Documentation*, **22**(2). June, 1966. 93–115.
General

ROLLING, L. N. *and* PIETTE, J. Interaction of economics automation in a large-scale retrieval system. *In:* Samuelson, K., ed. Proceedings of the FID/IFIP Joint Conference on Mechanized storage, retrieval and dissemination, Rome, 1967. Amsterdam: North Holland Publishing Co., 1968. 367–390.
Economics

SALMON, S. R. Library automation systems. New York: Dekker, 1975. 314 p. ISBN 0 8247 6358 0.
General

SALTON, G. Automatic information organization and retrieval. New York, London: McGraw-Hill, 1968. 522 p.
General

SALTON, G. ed. Information storage and retrieval: reports on indexing theory, content analysis, feedback searching and dynamic document space. Ithaca: Cornell University Department of Computer Science, 1974, var. pag. National Science Foundation, Scientific report no. ISR-22.
General

SAMS, H. W. & CO., INC., *comp.* Foulsham-Sams pocket dictionary of computer terms. Slough: Foulsham, 1965. 96 p.
Computing terminology

SAMUELSON, K., ed. Proceedings of the FID/IFIP Joint Conference, Mechanized information storage, retrieval and dissemination, Rome, 1967. Amsterdam: North Holland Publishing Co., 1968. 743 p.
General, essential

SAVAGE, T. R. The interpretation of SDI data. *American Documentation*, 18(4). October, 1967. 242–246.
Evaluation

SCHECHTER, G., ed. Information retrieval, a critical view: based on 3rd Annual Colloquium on Information Retrieval, May 12–13, 1966, Philadelphia. Washington: Thompson; London: Academic Press, 1967. 296 p.
General

SHARP, J. R. Some fundamentals of information retrieval. London: Deutsch, 1965. 224 p.
General

SIMS, D. M. What is a system analyst? *Special Libraries*, 59(9). November, 1968. 718–721.
System analysis

SPROWLS, R. C. Computers: a programming problem approach. New York, London: Harper and Row, 1969. rev. ed.
Programming

SWETS, J. A. Effectiveness of information retrieval methods. *American Documentation*, 20(1). January, 1969. 72–89.
Evaluation

Index

Index

This index does not cover subjects mentioned in the Glossary.

access, 40
accessing records, 31, 40–45
addressing a field, 34–9
air-line ticket reservation systems, 12
ALGOL, 98
analysis, system, as distinct from system design, 2
analyst,
  relation to information staff, 74
  role of, 28, 30
  and designer, 26
assembler, 97, 98
availability of input, 22
awareness, current, system elements of, 1

"between", 48, 50, 56
Boolean operators, 46–60
brackets in enquiry expressions, 47
British Standard (flowcharting), 78

change in system, 87
characteristics of information, 3
choice of storage medium, 45
citation searching, 63
class numbers, 7
clerical work, 81
  cost of, 82
cluster(ing), 4
COBOL, 99
communication, formal lines of, 81
computer errors and failures, 88
computer/user interaction, 71, 91–101
concept groups in enquiries, 47
conceptual and quantitative information, compared, 8
connection tables as precision devices, 64
context used for precision, 66
costs, costing, 76, 80
cost of clerical work, 82
current awareness, 1, 6–7

data,
  base, definition of, 2
  elements of, as input, 22
  use of, choice of storage, 45
  test, 87
date, field or fields, 32
dedicated (word, bit), 53–4
descriptor,
  descriptor-page dictionary, 60
  ratio of, to entities, 3
  selection of, for an enquiry, 69–71
desk-top retrieval tools, 7
demands on system, 7
derived information systems, 11
design stage, 86
designer,
  and influence of, economic factors, 6
  relation of, to information staff, 74
  and analyst, 26
document, distinguished from information, 4

economic factors, influence on design of, 6
enquiry,
  formulation of, 13, 46–9
  logic of, 47
  prompting of, 69
  selection of descriptors for, 69–71
ephemeral systems, 12
equipment as a system constraint, 17, 85
errors, computer, 88
error conditions, 83
estimation of response, 70

facsimile (fax) transmission, 11
flowcharts, -ing, 48–9, 58, 75–9, 86
  symbols for, 77
fluctuating systems, 12
field, types of, 31–8
fields and sub-fields, 32
file,
  housekeeping, 40, 100
  organization of, 39–45
  security of, 87

financial constraints, 7
fixed field, 34
forecasting manpower requirement, 85
FORTRAN, 97

generic-specific relationship, 44, 66
goals, statement of, 6, 7, 16, 17, 85
group discussions, 28–30
group profiles, 7
Guide, Users', 80, 84

handover, 87
hash-coding, 43–4
high-level languages, 98
homographs, 62
housekeeping, 40, 100
human and machine limitations, 76

implementation,
  staff time required for, 83
  time table of, 80
index language, effect of changes in, 83
index-production programs, 7
index sequential filing, 40
indicator, 41
inferred information systems, 11
information, types of, 3–9
information staff, 15, 17, 28, 74, 80, 82–5, 88–9
initializing, 93
input, 7, 20–22
instruction, program, 91
invariant field, 35
inverted files, 43, 53–6, 68

languages, computer, 97–9
level, *see* planes
links, linking, 42, 44, 64
list processing, 42–3, 57–9, 68
logical operators, 46–60
look-up systems, 5
low-level languages, 97–9

machine limitations on system, 76
maintenance of system and programs, 87
management, 17, 18, 81–2
manpower requirements, 76, 85
manual operations shown in flowchart, 79, 86
manual vs. computer system, 16
manual, instructional, for operators, 79, 80, 87
  see also Users' Guide
merge/sort operations, 45, 71
microfilm, 11
multi-programming, 100–101

natural language searching, 61–3

new system run simultaneously with old, 87
novel facilities, 17
numerical information, 9

on-line searching, 71, 100–101
'Out Of' search, 49–50, 56

package programs, 13
page filing, 59–60
parameters, parameterizing, 94–6
peripherals, speeds of, 108
PLAN, 97
planes (levels) of enquiry logic, 47
pointers to fields/records, 41–2
precision, 64–6
preliminary system agreement, 80, 81, 86
probability of response, 70
program/programme, distinction between, 2
programs, programming,
  definition of, 91
  instruction of, 91
  language for, 91, 97–9
  maintenance of, 87
  package, 13
prompting in enquiry formulation, 69

read-write head, 60
read-write ring, 87
'real time' operation, 71, 100–101
recall, 64
records, 20–25, 40
repeating fields, 33–44
requirements,
  possible future, 25
  statement of, 7, 16, 18, 86
retrospective search, definition of, 6
role indicators, 64

SDI (Selective Dissemination of Information), definition of, 6
search techniques, 48–63
security of files, 87
selection of enquiry descriptors, 69–71
semi-compiled programs, 98
sentinel, 40
serial filing, 40, 49–53, 68
sets (in searching), 47
slab, 39
software, 99
sort/merge operations, 45, 71
source language, 97
speeds of peripherals, 110
staff relations, 74
staff time (in implementation), 83
statement of requirements, 16, 18, 86
static systems, 12

storage, 45, 56
sub-field, 32
system,
 agreement, 85–6
 analyst, 30, 81
 definition, 1
 design, 13–14, 86
 designer, 15, 81
 study, 20, 25–8
 testing and maintenance, 87

term dictionary, 63
test data, 87
test material, 82

time-table for implementation, 80
truncation of words, 66

UDC codes, searching on, 66
user/computer interaction, 71
Users' Guide, 79, 80, 84

variables, 93
variant field, 35

weighting, 65
'wild-bird' sessions, 28–30
word truncation, 66

201703